Bioeconomic Modelling and Fisheries Management

Bioeconomic Modelling and Fisheries Management

COLIN W. CLARK

Department of Mathematics
University of British Columbia

A Wiley-Interscience Publication

JOHN WILEY & SONS

New York • Chichester • Brisbane • Toronto • Singapore

Copyright © 1985 by John Wiley & Sons, Inc.

All rights reserved. Published simultaneously in Canada.

Reproduction or translation of any part of this work
beyond that permitted by Section 107 or 108 of the
1976 United States Copyright Act without the permission
of the copyright owner is unlawful. Requests for
permission or further information should be addressed to
the Permissions Department, John Wiley & Sons, Inc.

Library of Congress Cataloging in Publication Data:

Clark, Colin Whitcomb, 1931–
  Bioeconomic modelling and fisheries management.

  "A Wiley-Interscience publication."
  Bibliography: p.
  Includes index.
  1. Fishery management—Mathematical models.
2. Biology, Economic—Mathematical models. I. Title.

SH331.5.M48C57 1985   639′.2′068   84-23445
ISBN 0-471-87394-2

Printed in the United States of America

10 9 8 7 6 5 4 3 2 1

Preface

The title of this book—*Bioeconomic Modelling and Fisheries Management*—is subject to a variety of interpretations. In order to avoid misunderstandings, therefore, I shall take this opportunity to define briefly what I mean by the words.

First of all, the *fisheries* to be discussed in this book are almost exclusively commercial marine fisheries. The models discussed and the general results obtained also apply to freshwater commercial fisheries, but only a few such examples are mentioned explicitly. Taxonomically, the word *fisheries* is commonly applied to the exploitation of species other than fish, including marine mammals, crustaceans, sharks, squids, and other organisms, and I also follow this practice.

The term *management* will be used in a broad sense, covering biologically oriented regulations such as catch quotas, mesh restrictions, seasonal closures, and the like, as well as possible economic instruments such as royalties, taxes, vessel fees or subsidies, as well as quota allocation systems. Clearly the scope of management is very much dependent upon jurisdictional features. Thus the advent of 200-mile zones of Extended Fisheries Jurisdiction has greatly increased the interest in more intensive forms of management than were formerly considered feasible.

The term *bioeconomics* encompasses the interrelations between the economic forces affecting the fishing industry and the biological factors that determine the production and supply of fish in the sea. Paramount among bioeconomic phenomena is the influence that the discounting of future revenues has on incentives for resource conservation. It is a fundamental principle of resource economics that the higher the rate of discounting used by exploiters the lower the degree of conservation. This principle is particularly relevant to fisheries, in which the lack of resource ownership forces exploiters to adopt in essence an *infinite* discount rate. This is one way—not necessarily the only way—to understand the "overfishing problem" or, alternatively, the "tragedy of the commons" in fisheries.

Many other bioeconomic relationships exist. Of particular interest are the ways in which various methods of fishery management affect the economic performance of the fishing industry. For example, the management approaches that have traditionally been used have led almost inexorably to severe overcapitalization of fishing fleets and processing plants. While this result can easily be predicted on theoretical grounds, it appears that such bioeconomic relationships are either not understood or not considered important by many fisheries scientists and administrators. It is widely agreed that fishery management has often failed to achieve its objectives, but argument continues over the reasons behind this failure. Hardly ever does the discussion fall back on general bioeconomic principles (and when it does so, it often invokes erroneous principles that have been outdated for a quarter-century).

Finally, a word about the types of *models* presented in this work. The world of modellers has nowadays split into two opposing camps: the "simplistic" versus the "holistic" modellers. The simplistic modeller attempts to understand complex systems by constructing many simple models of isolated aspects of the system, from which he hopes to derive general principles and understanding both of the system under investigation and of other, similar systems. This methodology resembles the approach traditionally used in science.

The holistic modeller, on the other hand, attempts to construct models that represent as closely as possible the entire system. Such a model, possible only with the aid of the modern computer, allows the modeller to "experiment" by manipulating his model in ways that would not be feasible with the real system.

Each modelling approach has its advantages and disadvantages. The simplistic approach leads to the formulation of general theories but may require familiarity with advanced mathematical techniques. The assumptions adopted can be clearly identified and kept in mind when applying the results. If, however, the whole is really greater than the sum of its parts, then simplistic models may fail to reflect important features of the system.

The holistic approach allows for the modelling of complexity itself, but there is always the danger that the model may become as difficult to understand as the original system. The underlying assumptions are seldom transparent to anyone but the model's builder. Model output—reams of computer printout—is also usually incomprehensible to outsiders, besides being unpublishable. Finally, successful modelling of one system may provide very little insight into the characteristics of other systems.

In short, simplistic modelling corresponds to "open" scientific investigation, with assumptions, deductions, and conclusions available to scrutiny and criticism by all who have the necessary mathematical background, whereas holistic modelling is much more a "closed" kind of investigation, in which outsiders must rely on the assumptions, explanations, and conclusions supplied by the model builders.

The simplistic modelling approach is taken throughout this work, although the computer has often been used to obtain numerical solutions of mathematical equations. The implication for the reader is that, like any scientist, he must practice patience as the theory develops and not hope to have everything discussed at once. Particularly in an interdisciplinary subject such as ours, this requires discipline and careful attention to the details of the models and arguments presented. But at least everything is out on the table and not bound up in unavailable computer programs and printouts.

The mathematical background required of the reader is minimal, consisting of facility with the calculus and simple differential equations and, in Chapter 6, elementary probability theory. The optimization techniques of dynamic programming and optimal control theory, which are used extensively, are described briefly within the text. The diligent reader, however, may wish to consult the references. My earlier book *Mathematical Bioeconomics* (Wiley–Interscience, 1976) gives considerably greater detail concerning these methods and their application to resource economics.

COLIN W. CLARK

Vancouver, British Columbia
January 1985

Acknowledgments

My first debt is to my students, Bill Reed, Marc Mangel, and Tony Charles, who taught me much of what I now know while themselves learning a bit of what I once knew. They read parts of the manuscript and provided extremely valuable advice. So did Charles Hall, Viggo, Don Ludwig, and Brian Rothschild. The influence of Gordon Munro is evident throughout. Finally, my deep gratitude is due to my wife Janet, who processed every word several times over.

C.W.C.

Contents

1   Introduction

1.1   The Overfishing Problem

In the 1950s a Canadian economist, H. S. Gordon, was asked by federal fisheries authorities to provide an economic analysis of the persistent problem of low income among Canada's maritime fishermen. Gordon's theory of the "common property" fishery (Gordon 1954), which has since become a classic, not only explained the low income of fishermen, but also clarified in economic terms the so-called overfishing problem (Graham 1952). It explained how *economic overfishing* would be expected to occur in any unregulated fishery, while *biological overfishing* would occur whenever price/cost ratios were sufficiently high. Gordon's model also suggested possible solutions to the overfishing problem, and these have formed the basis number of "limited entry" programs which have recently been introduced in various countries. Both experience and more detailed bioeconomic modelling, however, now indicate that many of the remedies originally proposed on the basis of Gordon's model were far too simplistic to successfully overcome the "tragedy of the commons" (Hardin 1968) in fisheries.

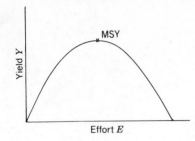

FIGURE 1.1 The sustainable yield–effort curve. (From Schaefer 1954.)

Because it illustrates both the advantages and limitations of simple bioeconomic models, we will begin by discussing Gordon's model in its original form. In the next section some of the main shortcomings of the model will be removed by extending it to a dynamic version. A case study requiring a somewhat different model is discussed in Section 1.3. The dynamic version of Gordon's model will serve as the basis for many of the more complex bioeconomic models to be discussed in this book.

Gordon's analysis was based, implicitly but not explicitly, upon a simple yield–effort curve employed by fishery biologists (Schaefer 1954)—see Figure 1.1. This curve depicts yield or annual catch Y as a function of fishing effort E. The latter term refers to the level of fishing intensity exerted on a given stock of fish by the fishermen. Thus, for example, fishing effort in a trawl fishery is normally specified in terms of the number of hours trawling (by standard trawlers) per annum. The concept of fishing effort, which is of fundamental importance in fishery management, will be discussed in considerable detail in Chapter 2.

"Yield" in Figure 1.1 refers to the annual catch that can be sustained over the long run if a fixed level of effort E is maintained. Thus "yield" really refers to *sustainable yield*, and Figure 1.1 provides a static or equilibrium view of the fishery. This is a serious limitation to the Gordon model, as will become apparent later.

The Schaefer yield curve reaches a maximum, referred to as the *maximum sustainable yield* (MSY), and then declines as effort is further increased. At sufficiently high sustained effort levels, yield falls to zero. *Biological overfishing* is said to occur whenever sustainable yield falls below MSY. The biological explanation for such an outcome, of course, is that intensive exploitation reduces the fish stock to a level at which its productivity begins to decline. (These processes are modelled in greater detail in the dynamic version of the Schaefer model discussed in the following section.)

To complete the Gordon model, we now let p denote the price of fish received by the fishermen and let c denote the costs of applying 1 unit of fishing effort. Both p and c are assumed here to be constant.

Before proceeding further, it is important to note that the word *cost* here is taken in the sense of *opportunity cost*, as normally understood in economic analysis. The opportunity cost of undertaking any activity is defined as the total cost of resources (e.g., labor and capital) employed in pursuing the activity, as determined by the most remunerative alternative uses to which these resources could have been put. The opportunity cost of fishing effort would thus include the actual costs of fuel, gear, and supplies, interest and depreciation on capital, as well as wages of skipper and crew. In particular, *wages* would mean the level of wages that fishermen could expect to earn in alternative employment opportunities rather than the actual wages (or shares) received by fishermen in a given season.

Figure 1.2 now completes the Gordon model. The curves labelled "Revenue" (R) have the equation $R = pY$, where Y is the yield curve of Figure 1.1. The straight lines labelled "Cost" (C) have the equation $C = cE$, cost being assumed to be proportional to fishing effort. The two cases depicted correspond to high- and low-cost parameters c, respectively.

The main prediction of Gordon's model can now be stated as follows. In an unregulated ("open-access" or "common property") fishery, effort will expand to the level $\bar{E}$ at which total revenue equals total cost. This level of effort represents an equilibrium, in which economic forces affecting fishermen and forces of biological productivity of the resource

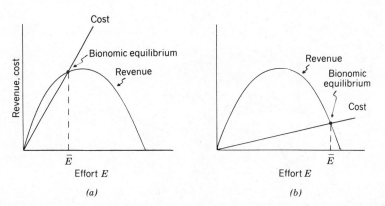

FIGURE 1.2 The static Gordon–Schaefer model: bionomic equilibrium at (a) high cost, (b) low cost.

are in balance. Gordon refers to this as the *bionomic equilibrium* of the common-property fishery.

A simple argument supports the prediction of bionomic equilibrium. First, clearly effort cannot exceed the level $\bar{E}$ in the long run, because then fishermen's opportunity costs would exceed their revenues, and some fishermen (at least) would take up alternative employment. Conversely, in the absence of entry restriction, a level of effort below $\bar{E}$ would induce additional fishermen to enter the fishery, attracted by revenues greater than what they could achieve elsewhere. Thus a (stable) equilibrium is established at $E = \bar{E}$.

The actual position of $\bar{E}$ is determined by relative prices and costs. Figure 1.2(*a*) shows the case in which fishing costs are relatively high. (Indeed, if costs were much higher than shown here, we would have $pY < cE$ for all values of E, and fishing would not develop at all.) On the other hand, Figure 1.2(*b*) shows the case where costs are low—or more generally, where the price/cost ratio p/c is relatively high. The more valuable the fishery, relative to the costs of fishing, the more intensively will it be exploited under open-access conditions.

It seems clear that the situation depicted in Figure 1.2(*b*) is an undesirable one. Effort has expanded to such a level that the resource stock has become depleted, and sustainable catches are reduced to a level far below the maximum possible. By reducing effort appropriately, one could accomplish a double improvement—catches (and hence fishermen's revenues) would increase, and the costs of fishing would decrease. From the fishermen's point of view it would appear that the optimum level of effort occurs at E^*, where the difference between revenues and costs is maximized (see Fig. 1.3).

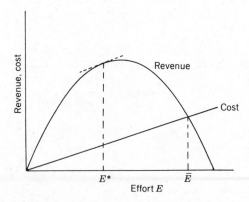

FIGURE 1.3 The static Gordon–Schaefer model: maximum economic yield (MEY) occurs at $E = E^*$.

Indeed, Gordon (1954) argues that both the fishermen's and the social welfare optimum occur at this point E^*, where the "economic rent" (the difference between revenue and cost) is maximized. In the literature E^* is often referred to as the position of *maximum economic yield* (MEY).

We are now in a position to raise a number of important questions pertaining to the above model and analysis:

1. Can the prediction of bionomic equilibrium be tested empirically?

2. Given that fishermen would be better off at the MEY position than at bionomic equilibrium, do we observe strong desires or attempts on the part of fishermen to improve their position? If not, why not?

3. Is government intervention required to improve the situation, and if so, what policies and objectives would be most appropriate?

4. More precisely, are the gains to society from fishery regulation sufficiently large to warrant the cost of an effective management program?

Of course, many other questions may be raised pertaining to both the biological and economic verisimilitude of the Gordon model. Such questions, which may be quite important in practical situations, will be addressed later in the book. For the moment, however, let us continue to work within the present framework. We will deal briefly with the four questions raised above.

Empirical Verification

A rather extensive literature of economic case studies of commercial fisheries is now available (e.g., Christy and Scott 1965, Crutchfield and Pontecorvo 1969, Bell 1978, Neher and Scott 1980). Works such as these lend rigorous academic support to Gordon's analysis of the unregulated open-access fishery. But anyone familiar with the worldwide history of fishery development since the 1950s has no need of academic verification to realize that Gordon's predictions have by now been borne out over and over again. Depletion of major fish stocks and the impoverishment of fishing fleets and processing companies have become common phenomena worldwide.

Several major fisheries have collapsed completely (see Table 1.1), largely as a result of overfishing—although in many cases environmental

TABLE 1.1   Some Depleted Marine Resource Stocks

Stock	Peak Catch (year)	1981 Catch	Reference
Antarctic blue whales	29,000 whales (1931)	Nil	FAO[a] (1979)
Antarctic fin whales	27,000 whales (1938)	Nil	FAO[a] (1979)
Hokkaido herring	850,000 tons (1913)	Nil	Murphy (1977)
Peruvian anchoveta	12.3 million tons (1970)	0.3 million tons	IMARPE[b] (1974)
Southwest African pilchard	1.4 million tons (1968)	Nil	Butterworth (1980)
North Sea herring	1.5 million tons (1962)	Negligible	Saville (1980)
California sardine	640,000 tons (1936)	Nil	Murphy (1977)
Georges Bank herring	374,000 tons (1968)	Nil	Sinderman (1979)
Japanese sardine	2.3 million tons (1939)	17,000 tons (1973)	Murphy (1977)

[a]United Nations Food and Agriculture Organization.
[b]Institut del Mar del Peru.

influences may have also played a role in these collapses. Many other fisheries are currently yielding catches well below both historical and estimated MSY levels. And even where high catch rates have been maintained, excessive capacity of fishing fleets and processing plants is everywhere the rule rather than the exception.

Fisheries biologists are well aware that many fish populations undergo large-scale natural fluctuations (see, e.g., Sissenwine 1984). The development of a commercial fishery may be first precipitated when a population is at a peak of abundance. If the stock subsequently declines, it may be difficult to disentangle the influences of fishing pressure and natural processes. The pelagic schooling species (herrings, anchovies, and like species) are particularly subject to cycles of abundance and decline and are easily overfished. It is clear that the uncontrolled growth of a large fishing industry has the potential for hastening and exacerbating the decline, as well as preventing the recovery of such stocks.

The Gordon model, because of its particular functional form (see Section 1.2), in fact predicts that

$$\bar{E} = 2E^*$$

that is, that the bionomic equilibrium level of effort will equal exactly twice the optimum level. (In this sense, *economic overfishing* always occurs in the open-access fishery.) In practice, however, it often appears that the effort *capacity* of fishing fleets is much larger than twice the optimum level; an example where the current fleet size seems to be nearly 10 times the optimum level is discussed in Section 1.3.

If anything, the Gordon model probably underestimates the degree of overexpansion and stock depletion possible in fisheries. For example, as explained in Section 1.2, there is good reason to expect that the open-access fishery will initially expand *beyond* the long-run bionomic equilibrium level, and this expansion may be long-lasting. Also, the depletion of the resource may, in extreme cases, be irreversible once a critical threshold is passed.

Is MEY Optimal?

The benefits from reducing effort in an open-access fishery appear so obvious that some authors have referred to this action as a "no losers" option. But this view fails to explain why rational fishery management has historically proved to be so difficult to achieve.

One aspect of the problem that does not show up in the static Gordon model is this: *It is never possible to move instantaneously from bionomic equilibrium to an "improved" position such as MEY*. Such a shift requires that the fish stock first be rehabilitated to more productive levels, and stock rehabilitation always takes time to accomplish. Frequently the necessary recovery period will be quite protracted. For example, it has been suggested that many stocks of Pacific salmon are currently depleted (Pearse 1981). But reproductive cycles in these populations take from three to five years. Reductions in current harvest levels would thus only lead to increased recruitment after several years' delay. In the case of severe depletion, several decades may be required before stock increases become apparent. Antarctic blue whale stocks, for example (Table 1.1), will probably take at least 50 years to recover to the MSY level, assuming that the present moratorium on this species remains effective.

In other words, a reduction in fishing effort will initially result in a *reduction* in catches and fishermen's revenues, rather than in increased catches and revenues, as might be suggested by the simple Gordon model. Such increases can only occur following a period of stock rehabilitation.

The extent to which current losses (or potential losses) of income are justified by future gains is an important and central problem in economics. But this problem cannot be addressed on the basis of a static equilibrium model. The dynamic version of the Gordon–Schaefer model to be discussed in the next section will in fact demonstrate that MEY is unlikely to be a desirable objective, either to the fishing industry or to society as a whole. In fact, the dynamic optimum will be seen to involve a certain "compromise" between the MEY and bionomic equilibrium positions.

We see, from this discussion, that Gordon's bioeconomic model, successful as it may be in explaining the overfishing phenomenon, appears inadequate as a model of economically desirable management policy. The static or equilibrium model is too simplistic to provide reasonable predictions of the behavior of resource exploiters or to serve as a reliable guide to socially desirable management policy. Even more important, the static model can also be extremely misleading if used to predict the actual effects of various alternative management programs. Let us see why this is so.

Government Management Policy

Let us now imagine a typical situation, in which an unregulated fishery has reached bionomic equilibrium. Catches are low, fishermen are impoverished, and "government action" is demanded. In order to reduce fishing the government initiates a program of total catch quotas (or total allowable catches—TACs). When the TAC for a given year has been taken, the fishery is closed until the following year.

At first the TAC must be set at a low level, to permit stock rehabilitation. Ultimately the stock does recover to the optimum MEY level, say, and the TAC is then raised to the MEY catch. The ideal situation depicted in Figure 1.3 has apparently been reached. Catches have increased and are now obtained with less fishing effort from a more abundant stock.

Now we return to Gordon's argument. Suppose indeed that—by means of closures, for example—fishing effort is maintained at the level E^*. At the fixed effort level E^* large profits accrue to fishermen. Additional fishermen are therefore attracted to the fishery, and fishing *capacity* consequently increases. Unless this new entry is prevented, the authorities will be forced again to shorten the fishing season in order to maintain catches at the TAC level. The MEY catch is now being shared among

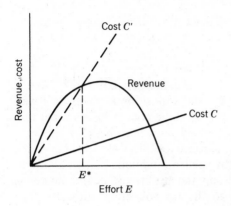

FIGURE 1.4 Adjustment in the cost curve in an open-access, catch-regulated fishery (see text).

more fishermen, and the *costs* of fishing have increased (since vessels are idle, or employed in less lucrative fisheries, during the closed season). An equilibrium can be reached only when the cost curve has been pushed up to a new position *C'*, as shown in Figure 1.4. In fact, it can be seen (see Chapter 3) that the ultimate fishing capacity will be *greater* than it would have been at bionomic equilibrium in the absence of any regulation. The effect of regulation has simply been to replace overfishing with over-capacity.

Such results, which may seem counterintuitive at first, suggest that caution is needed in the use of simplistic bioeconomic models as guides to management policy. Thus the "pictorial" model of Figure 1.3 seems to say that controlling "effort" at an appropriate level is all that is required for efficient operation of the fishery. A more careful analysis, however, shows that such policies may have the pernicious effect of actually *increasing* the overexpansion of capacity in the fishery.

Indeed, the overcapacity problem has now become severe in many managed fisheries. For example, after years of governmental attempts to control effort in Canada's Pacific fisheries, a federal commissioner's report (Pearse 1981) stated:

The central problem for all the commercial fisheries is the fleets' chronic overcapacity. All of our major fisheries, especially the salmon, herring, and halibut fisheries, have greatly expanded their fishing power in recent years. But because the quantity of fish has not increased, most of the new capital investment in vessels and gear and the advanced technology that has been added to the fleets is wasted. Our most valuable stocks could be fully harvested with only a fraction of the capital and labour now expended on fishing them. This

*wasteful pattern of development reflects governments' failure, in spite
of repeated attempts, to develop a policy that would encourage the
industry to develop efficiently.*

An alternative approach, often discussed theoretically but seldom if
ever implemented in practice, involves the imposition of *taxes*, or royalties. For example, suppose fishermen were to be charged a tax τ per unit
of effort. The perceived cost curve would then become $C' = (c + \tau)E$,
and this could be adjusted to any desired position—for example, the
position C' shown in Figure 1.4. In such a situation, bionomic equilibrium would coincide with MEY, and the profits (or economic rents)
from the fishery would all accrue to the tax-collecting authority. The
reader will readily perceive why management via taxes holds little appeal
to the fishermen. (Interestingly, tax schemes have upon occasion even
been recommended for *international* fisheries, the nonexistence of international tax-collecting authorities having been apparently overlooked!)
Another approach, which in theory (at least) is equivalent to taxes in
terms of economic efficiency, is the use of *exclusive fishing rights*. Such
rights can be allocated to individual fishermen, vessel owners, fishing
companies, or a combination of these. Benefits accrue to the rights
owners. These methods, which are currently being considered in several
countries, also involve serious problems (monitoring and enforcement,
for example). Further discussion is deferred until Chapter 4.

What does seem clear at this stage is that fishery systems are much too
complex—biologically, socially, and economically—for there to exist any
simple solution to the problems of fishery management. Unfortunately,
the various "actors" in any particular fishery often tend to take somewhat
myopic views of the system, with the result that the overall management
problem never becomes fully articulated (United Nations FAO 1980).
Perhaps the present book will help to bring the various viewpoints into
perspective.

Benefits and Costs of Management

Given that fishery systems are complex, it follows that effective
management may be expensive. Returning to Figure 1.2, note that the
distance separating the revenue and cost curves provides an indication of
the potential economic benefits to be obtained from management. These
can be large or small. It may seem trite to observe that a complex
management program will only be justified if the potential benefits are
significant. However, the economic benefits of many existing management programs are at best dubious. Some commentators (e.g., Christy

1977) have even asserted that the net returns to society from fisheries management have so far probably been negative. It is a well-known sociological law that regulatory bodies tend to become manipulated by the industry they were set up to regulate (Niskanen 1971), but the fishing industry's apparent lack of concern over its own long-term welfare remains hard to explain, except perhaps on the basis of a real misunderstanding of the bioeconomic system.

N.1

Efficiency vs. Employment

An issue that is often raised in discussions of efficiency in fishery management is the problem of unemployment among fishermen. If fishing capacity is to be reduced, or limited to the optimal level, some fishermen may no longer be employed in the fishery. Unless the fishermen are able to find suitable alternative employment elsewhere, this reduction in employment may be a serious social problem. But otherwise, transferring surplus fishermen, whose employment in the fishery adds nothing to production, to other employment, whether in an alternative fishery or elsewhere, will constitute a net benefit to society as a whole.

Labor, however, is not the only input to fishing—except for artisanal fisheries. Unless inputs are regulated, the open-access fishery will experience a surplus of both capital and labor. Moreover, the type of capital introduced may be excessively "efficient" in terms of individual vessel operation, as each vessel owner attempts to outcompete his competitors for the limited catch available—a completely antiproductive competition.

In economic terminology, the social opportunity cost of labor will be zero if structural unemployment exists in the fishing community. The opportunity cost of capital, however, is almost never zero, either in developed or underdeveloped countries. (But the opportunity cost of *emplaced* capital may well be zero—see Sec. 3.3.) If fisheries are to be managed for socio-economic objectives, these complications must be faced.

1.2   A Dynamic Model

We now discuss a dynamic bioeconomic model, for which the Gordon model of Section 1 is the equilibrium solution. The fact that the new model is still woefully oversimplified is of course irrelevant at this initial stage. (When teaching this material I ask students to formulate at least 10 major criticisms of the basic model. The class usually has no difficulty coming up with about 25 criticisms.)

The biological component of our dynamic model is due to Schaefer (1954), who applied it to the tuna fisheries of the tropical Pacific. The model is based on the *logistic* population model:

$$\frac{dX_t}{dt} = rX_t\left(1 - \frac{X_t}{K}\right) \tag{1.1}$$

Here X_t denotes the fish *population biomass*, dX_t/dt is the time rate of change of X_t, and r and K are positive constants referred to as the *intrinsic growth rate* and the *environmental carrying capacity*, respectively. The reason for these terms is simple: r represents the maximum relative growth rate of the population, which is the approximate rate of (exponential) growth when $X_t \ll K$. Similarly, K represents the stable equilibrium biomass level: If X_t is a positive solution of Eq. (1.1), then $X_t \rightarrow K$ as $t \rightarrow \infty$. A detailed discussion of this model and the ecological significance of the parameters r and K are given in May (1981).

The expression $G(X) = rX(1 - X/K)$ on the right side of Eq. (1.1) of course represents the *natural growth rate* of the population. To model the effect of fishing on population dynamics, Schaefer altered the equation to

$$\frac{dX_t}{dt} = rX_t\left(1 - \frac{X_t}{K}\right) - C_t \tag{1.2}$$

where C_t denotes the *catch rate*. Finally, Schaefer expressed catch in terms of *effort* by the simple relation

$$C_t = qE_tX_t \tag{1.3}$$

where E_t denotes fishing effort and q is a constant called the *catchability coefficient*.

Fishing Mortality

We define *fishing mortality* F_t as the relative mortality rate due to fishing

$$F_t = \frac{C_t}{X_t} \tag{1.4}$$

Thus

$$C_t = F_tX_t \tag{.5}$$

where, by Eq. (1.3),

$$F_t = qE_t \tag{1.6}$$

It is important to realize that Eq. (1.5) is merely a *definition* (of fishing mortality), whereas Eq. (1.6) constitutes an important *hypothesis* of the Schaefer model. Specifically, this hypothesis states that fishing mortality is directly proportional to fishing effort. This hypothesis derives from a certain probabilistic model of the fishing process; see Chapter 2, where other possibilities will be discussed.

When rewritten in the form

$$\frac{C_t}{E_t} = qX_t \tag{1.7}$$

the Schaefer hypothesis asserts that *catch per unit effort* (CPUE) *is a direct (proportional) index of stock abundance X*. Since X is usually not observable, while catch and effort data can be obtained from the fishery, Eq. (1.7) provides a valuable approach to stock assessment. Two important limitations, however, are that (1) the value of the catchability coefficient q cannot be estimated unless some independent assessment of stock size is available, and (2) the basic Eq. (1.7) is only a hypothesis, and as such may result in biased estimates. This hypothesis is now in fact thought to be seriously misleading, especially for pelagic schooling species of fish (see Chapter 2 for further discussion).

The Static Model

The static yield–effort curve of Section 1.1 consists of the equilibrium solutions of the dynamic Schaefer model. Thus, setting $dX_t/dt = 0$ in Eq. (1.2), and copying (1.3) again, we have (omitting subscripts)

$$C = rX\left(1 - \frac{X}{K}\right) \quad \text{and} \quad C = qEX$$

These equations imply that

$$X = K\left(1 - \frac{qE}{r}\right) \tag{1.8}$$

and therefore

$$C = qKE\left(1 - \frac{qE}{r}\right) \tag{1.9}$$

Equation (1.9) thus represents the relationship between catch and effort *at equilibrium*. If effort is held constant at E, it can easily be shown (for our model) that the fish stock biomass will approach the equilibrium value given by (1.8), and catch will then reach the equilibrium of (1.9). If effort is changed to a new level, the system will adjust to a new equilibrium, at a rate depending upon the parameters.

Equation (1.9) generates the parabolic yield–effort curve shown above in Figure 1.1. Note that the MSY point on this curve is given by

$$E_{\text{MSY}} = \frac{r}{2q}, \qquad C_{\text{MSY}} = \frac{rK}{4} \qquad (1.10)$$

The static Schaefer curve (1.9) also provides a crude method for estimating MSY from catch-effort data. Namely, Eq. (1.9) implies a linear relation

$$\frac{C}{E} = a - bE$$

whose coefficients can be estimated by simple linear regression. MSY is then given by

$$C_{\text{MSY}} = \frac{a^2}{4b}$$

For more sophisticated statistical treatments, see Schaefer (1967), Schnute (1977), Ludwig and Hilborn (1982), and Cooke (1984).

The Dynamic Gordon–Schaefer Model

Let us now add Gordon's economic component to the dynamic Schaefer model. Assuming a constant price p for fish and a constant unit effort cost c, we obtain the expression

$$\pi_t = pC_t - cE_t$$
$$= (pqX_t - c)E_t \qquad (1.11)$$

for the *net revenue flow* to the fish harvesting industry. The *bionomic equilibrium* biomass level $\bar{X}$ is determined from the condition $\pi = 0$; that is,

$$\bar{X} = \frac{c}{pq} \qquad (1.12)$$

Thus, as expected, lower cost/price ratios result in proportionally more severe stock depletion under unregulated open-access exploitation. Note that $\bar{X} > 0$ for every possible parameter combination (except $c = 0$). Thus the Gordon–Schaefer model predicts that extinction of the fish population will not occur, the reason being simply that at sufficiently low

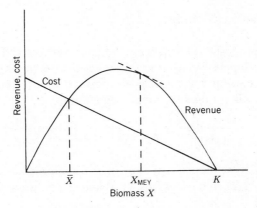

FIGURE 1.5 The static Gordon–Schaefer model: cost and revenue in terms of the biomass level X.

stock levels X the catch rate qEX becomes unprofitable (i.e., $\pi = pqEX - cE$ is negative for small X), and fishing therefore ceases. Actual extinction of formerly productive fish stocks seems in fact to be a rarity, but this "prediction" is of course a feature of the model, and not a basic bioeconomic principle, as has sometimes been asserted. The nonextinction prediction depends particularly on (1) the form of the assumed growth relationship $G(X)$ (see Clark 1976a, sec. 1.2), and (2) the form of the assumed catch expression qEX (see Chapter 2).

For later reference it will also be useful to express sustainable revenue and cost in terms of biomass X. The algebra, which is simple, results in

$$\text{Revenue} = pC = prX\left(1 - \frac{X}{K}\right) \tag{1.13}$$

$$\text{Cost} = cE = \frac{cr}{q}\left(1 - \frac{X}{K}\right) \tag{1.14}$$

These curves are shown in Figure 1.5, which also shows the bionomic equilibrium biomass $\bar{X}$ and the MEY point X_{MEY} (corresponding to E^* in Fig. 1.3).

Note again, in Figure 1.5, that once biological overfishing occurs (i.e., X falls below $X_{MSY} = K/2$), further decreases in the stock level result in both decreased sustained revenue and increased sustained costs. This is the sense in which overfishing appears to be doubly irrational—a somewhat misleading appearance, as we will soon see (following a short digression).

Smith's Open-Access Model

Smith (1969) proposed the following dynamic version of the open-access fishery model. Suppose that effort E represents the number of vessels in the fishery. If current revenues π_t are positive, E will tend to increase, and vice versa. Smith uses the simple equation

$$\frac{dE}{dt} = k\pi \qquad (k = \text{constant}) \tag{1.15}$$

to model fleet dynamics. Combining this equation with the population dynamics, Eq. (1.2), and using Eqs. (1.3) and (1.11), we obtain the system of equations

$$\left.\begin{aligned} \frac{dX}{dt} &= rX\left(1 - \frac{X}{K}\right) - qEX \\ \frac{dE}{dt} &= k(pqX - c)E \end{aligned}\right\} \tag{1.16}$$

This plane autonomous system of differential equations can be analyzed by the usual graphical method (Clark 1976a, chap. 6). The resulting trajectories are illustrated in Figure 1.6 [the "phase-plane diagram" of the system (1.16)]. It can be demonstrated that the bionomic equilibrium point $(X = \bar{X})$ is a *stable* equilibrium and that all trajectories

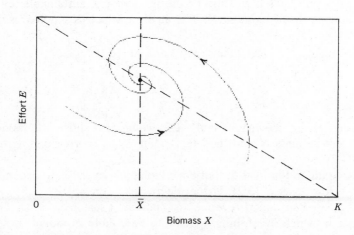

FIGURE 1.6 Trajectories of Smith's model of the open-access fishery. The bionomic equilibrium at $X = \bar{X}$ is reached in a cyclical fashion (see also Fig. 1.7).

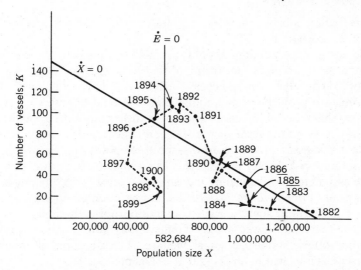

Figure 1.7 Development of the North Pacific fur seal fishery, 1882–1900 (From Wilen 1969).

converge to this equilibrium. However, oscillations may occur, as illustrated.

A fishery, for example, that develops on a lightly exploited stock may experience a "boom-and-bust" development phase, with large levels of fishing capacity resulting in stock depletion *beyond* Gordon's predicted equilibrium. Ultimately vessels withdraw from the unprofitable fishery, and the bionomic equilibrium is finally reached. (A more incisive analysis of this phenomenon results from considerations of fixed cost vs. variable cost—see Chapter 3.)

Boom-and-bust development histories are indeed common in fisheries, although other causes may also contribute [for example, environmentally induced stock fluctuations (Chapter 6) or uncertainty as to natural productivity parameters]. Wilen (1969) has found a good qualitative fit to Smith's model of data from the nineteenth-century and early-twentieth-century North Pacific fur seal fishery (Fig. 1.7).

The Sole-Owner Optimum

Economics textbooks often consider two extreme, or idealistic, forms of industrial organization: pure competition on the one hand, and pure monopoly on the other. While few actual industries fit either model closely, the idealistic models are relatively easy to formulate and to analyze. Oligopolistic theories, on the other hand, are notoriously difficult to work with.

The same classification applies to models in resource economics, although there are important distinctions. We have already examined the case of competitive, "open-access" fishing and have concluded that a long-run zero-profit bionomic equilibrium will tend to emerge. We now examine the case of a *sole owner* of a fishery resource stock. (We deliberately avoid the term *monopolist* here, because this term refers to a firm that possesses *market* control—the monopolist can fix prices. Our present Gordon–Schaefer model assumes no such market power—indeed, we are assuming that the fishing industry is a "price taker," facing a fixed and constant price level p. This distinction between monopolist and sole owner is important, since market competition normally improves social welfare, whereas unregulated competition for common-property resources often diminishes social welfare. The question of monopoly power in fisheries will be investigated in Chapter 3.)

We will hypothesize that the sole owner attempts to maximize his profits from the fishery. In a dynamic framework, however, the concept of "profit" needs some clarification. Let i denote the annual (*real*) rate of interest in the economy. (The real rate of interest is the noninflationary rate and equals the nominal rate of interest minus the rate of inflation.) Thus i represents the (real) annual rate of return that investors expect to earn on their investments.

A sum of \$$P$, invested at compound interest with interest rate i, will in t years' time be worth \$$P_t$, where

$$P_t = P(1 + i)^t$$

The *present value* of a payment P_t due in t years' time is defined as the sum P_0 that must be invested now to provide P_t in the future:

$$P_0 = \frac{P_t}{(1 + i)^t}$$

The factor $(1 + i)^{-t}$, called the *discount factor*, is more conveniently expressed in the exponential form $e^{-\delta t}$, with

$$(1 + i)^{-t} = e^{-\delta t}, \quad \text{or} \quad \delta = \ln(1 + i)$$

Here δ is called the annual (continuous) *discount rate*. Our basic discounting formula is then:

$$P_0 = P_t e^{-\delta t} \tag{1.17}$$

There are strong economic arguments to support the hypothesis that the investment decisions of firms are made with the objective of maximizing the wealth of the firm, as measured by the discounted present value of its net future income flow (see, e.g., Hicks 1946). This hypothesis, for example, underlies most of the current literature on exhaustible resource economics (Hotelling 1931, Dasgupta and Heal 1979).

Following this tradition, we write the present value of net revenue flow π_t, over a time interval of length dt, as $\pi_t e^{-\delta t} dt$. The *total present value* of π_t over the time horizon from $t = 0$ to $t = T$ is therefore given by

$$PV = \int_0^T \pi_t e^{-\delta t} dt$$

It is both convenient and natural to assume that $T = +\infty$ (otherwise undesirable "horizon effects" occur, in which assets are simply "cashed in" at $t = T$). Substituting from Eq. (1.11) for the fishery model, we finally formulate the sole owner's objective as

$$\text{Maximize} \int_0^\infty (pqX_t - c)E_t e^{-\delta t} dt \qquad (1.18)$$

The firm's only decision variable, in this formulation, is the amount of fishing effort E_t to be used at each instant $t \geq 0$. The stock biomass X_t for all future times $t \geq 0$ is then determined from our Schaefer model, Eqs. (1.2) and (1.3); the initial biomass X_0 is also assumed to be known.

To summarize, the sole owner of the fishery wishes to determine his effort schedule, E_t $(t \geq 0)$, so as to maximize the profit objective

$$\int_0^\infty e^{-\delta t}(pqX_t - c)E_t \, dt \qquad (1.19)$$

subject to the equation

$$\frac{dX_t}{dt} = G(X_t) - qE_tX_t, \qquad X_0 \text{ given} \qquad (1.20)$$

where E_t satisfies

$$E_t \geq 0 \qquad (1.21)$$

Here $G(X)$ denotes the natural growth rate of the fish biomass X; in the Schaefer model we have

$$G(X) = rX\left(1 - \frac{X}{K}\right)$$

but our theory applies equally to other growth functions $G(X)$, such as those studied by Pella and Tomlinson (1969), Shepherd (1982b), and others. The differential equation (1.20) is often called a *general production model*.

We will henceforth refer to the solution of the foregoing mathematical optimization problem as the *optimal* solution. No value judgment is attached to this purely technical use of the word *optimal*, but we will later discuss social implications of the profit-maximizing optimal harvest policy.

In mathematical terms, the above maximization problem belongs to the realm of the calculus of variations (or in modern terms, optimal control theory; see Clark 1976a, chap. 4). It happens, however, to be an unusually simple problem of this type, one that can be solved directly by elementary means. To keep the mathematics as simple as possible, we shall for the moment adopt the extra simplification of supposing $c = 0$. The general case $c \neq 0$ can be solved by the same method, and details are given in the Appendix to this chapter.

With $c = 0$, we now substitute for the expression qEX from Eq. (1.20) into (1.19), which becomes

$$\int_0^\infty e^{-\delta t} \left[G(X_t) - \frac{dX_t}{dt} \right] dt$$

Next we perform an integration by parts for the second term in the integrand:†

$$\int_0^\infty e^{-\delta t} \frac{dX_t}{dt} dt = \delta \int_0^\infty e^{-\delta t} X_t \, dt - X_0$$

Since X_0 is a given constant, our maximization problem can be expressed as

$$\text{Maximize} \int_0^\infty e^{-\delta t} [G(X) - \delta X] \, dt \tag{1.22}$$

But this expression is maximized provided that the expression $G(X) - \delta X$ is maximized. Thus the sole owner should employ a fishing policy that maintains the biomass at the level X which maximizes

$$G(X) - \delta X$$

By elementary calculus, the optimal biomass level $X = X^\#$ satisfies

$$G'(X^\#) = \delta \tag{1.23}$$

---

†The integration-by-parts formula is

$$\int_a^b u \frac{dv}{dt} dt = uv \Big|_a^b - \int_a^b v \frac{du}{dt} dt$$

If the initial biomass X_0 happens to be at $X^\#$, it should be kept there. Otherwise a fishing program should be used to shift X_t as rapidly as possible from X_0 to the optimal level $X^\#$ [since any delay in shifting reduces the value of (1.22)].

The optimal effort policy $E_t^\#$ of our (simplified) model can therefore be expressed as:

$$E_t^\# = \begin{cases} E_{max} & \text{if} \quad X_t > X^\# \\ \dfrac{G(X^\#)}{qX^\#} & \text{if} \quad X_t = X^\# \\ 0 & \text{if} \quad X_t < X^\# \end{cases} \qquad (1.24)$$

where E_{max} denotes the maximum effort capacity available to the owner.† Equations (1.24) simply say that the owner should fish at the maximum feasible rate if X_t is above the optimal level $X^\#$ and should refrain from fishing when X_t is below $X^\#$. Once X_t reaches $X^\#$ it should be kept there [$dX/dt = 0$ implies that $E = G(X)/qX$, by Eq. (1.20)]. A policy of this kind is called, colorfully, a "bang–bang" control policy in the literature.

Equation (1.23), that is, $G'(X^\#) = \delta$, is very well known in the economic theory of capital and investment, where it is sometimes called the "golden rule of capital accumulation." Indeed, if we consider the resource stock X as a capital asset (cf. Clark and Munro 1975), the analogy is complete. The function $G(X)$ can be interpreted as the natural rate of production of the resource stock X. Then $G'(X)$ represents *marginal productivity* of the resource, and Eq. (1.23) tells us that the present value of production output is maximized when marginal productivity of the resource asset equals the rate of discount.

In the special case of zero discounting, we have $G'(X^\#) = 0$, which implies that $X^\# = X_{MSY}$, the MSY biomass level. Thus MSY would be the sole owner's optimum policy provided he used a zero discount rate (and ignored the cost of fishing). When $\delta > 0$ we see that $X^\# < X_{MSY}$; the larger δ, the smaller the optimal biomass level $X^\#$. Although obtained here only under restrictive assumptions, this last rule is in fact of quite general validity in resource economics (and in capital theory in general). We state it as a *fundamental principle* of renewable resource economics:

Higher discount rates normally imply lower levels of resource conservation by private resource owners, other things being equal.

---

†Here we suppose E_{max} to be given and to satisfy $E_{max} > G(X^\#)/qX^\#$. The problem of determining the optimal effort capacity is interesting but is nontrivial in a dynamic setting; see Chapter 3.

An extreme manifestation of this rule arises in our model if the discount rate δ is larger than the intrinsic growth rate of the population $r = G'(0)$. For in this case we see that $G(X) - \delta X$ is maximized for $X = 0$—that is, the "optimal" profit-maximizing policy is the extinction of the resource! The resource owner, by assumption, has alternative uses for capital funds, which will provide him a greater rate of return (δ) than will the resource asset. Of course, this conclusion depends upon the formulation of our model. When effort costs are reintroduced into the Schaefer model, for example, extinction is never optimal, since fishing costs exceed revenues whenever X falls below $\bar{X} = c/pq$. For other aspects of the extinction question see Clark (1971, 1973a), Beddington, Watts, and Wright (1975), and Clark and Munro (1980).

In any case, our simple model does suggest that owners of *slowly growing* renewable resource stocks might have an economic incentive to deplete such stocks. Typical examples of resources of this kind include: whales and other large mammals (see below), sharks, large demersal fish species, temperate forests, desert savannahs, soils, and so forth. All these resources have been subject to serious depletion at various times and places. While it is obviously unreasonable to attach the blame to high discount rates in all such cases, there is no doubt whatever that the discounting of future returns, both by private firms and by society at large, can and does have serious anticonservationist implications. This fact has long been recognized in the resource conservation literature (Ciriacy-Wantrup 1968, Page 1976).

Cost Effects

We now reintroduce effort costs into our model. Write

$$c(X) = \frac{c}{qX} \tag{1.25}$$

It is easy to show that $c(X)$ represents the cost of catching one unit of fish biomass when the total biomass is X. (If catch $= 1 = qEX \, \Delta t$, then cost $= cE \, \Delta t = c/qX$.) Modifications of the Schaefer catch equation (1.3) lead to alternative forms for this unit cost function; see Section 2.2.

The sole owner's optimal fishing problem is described by Eqs. (1.19)–(1.21) above. Once again (as shown in the Appendix) there exists an *optimal biomass level* X^*, and the optimal fishing policy is the "bang-bang" approach to X^* given by Eqs. (1.24) (with $X^{\#}$ replaced by X^*). The value of X^* is determined by the fundamental equation

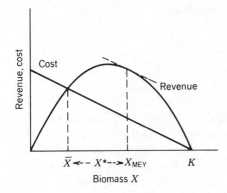

FIGURE 1.8 The optimal biomass X^* [Eq. (1.26)] lies between $\bar{X}$ and X_{MEY} and decreases from X_{MEY} toward $\bar{X}$ as δ increases from 0 to ∞.

$$G'(X^*) - \frac{c'(X^*)G(X^*)}{p - c(X^*)} = \delta \qquad (1.26)$$

which is just a modified form of Eq. (1.23), involving the additional term

$$\frac{-c'(X^*)G(X^*)}{p - c(X^*)} \qquad (1.27)$$

This term is called the *marginal stock effect term* (Clark and Munro 1975) since it specifies the effect that stock "thinning" has on the optimal biomass level: As the fish stock is thinned out (by fishing it down), the unit costs of harvesting increase, and (1.27) reflects this effect.

It will be shown in the Appendix that the optimal biomass X^* lies *between* the bionomic equilibrium $\bar{X}$ and the Gordon static optimum X_{MEY} (as discussed in Section 1.1)—see Figure 1.5. Furthermore X_{MEY} equals X^* for zero discounting ($\delta = 0$), and $\bar{X}$ is the limiting position of X^* as $\delta \to \infty$. Also, in accord with the fundamental principle stated above, X^* decreases as the discount rate increases (see Fig. 1.8).

These are intuitively appealing results. When the future is not discounted, then all time periods are equally important and MEY is the sole owner's optimal policy. At an infinitely large discount rate, on the other hand, the owner has no concerns for the future and therefore attempts to maximize short-term profits only. This is also exactly what the open-access fishery does (see Scott 1955).

In the event that $G(X)$ is the Schaefer logistic function $rX(1 - X/K)$, Eq. (1.26) reduces to a quadratic equation from which the following closed formula for X^* is readily obtained.

$$X^* = \frac{1}{4}\left\{\bar{X} + K\left(1 - \frac{\delta}{r}\right) + \sqrt{\left[\bar{X} + K\left(1 - \frac{\delta}{r}\right)\right]^2 + \frac{8K\bar{X}\delta}{r}}\right\} \qquad (1.28)$$

where, as before, $\bar{X} = c/pq$. Two special cases are

$$\delta = 0: \qquad X^* = \frac{1}{2}(\bar{X} + K) = X_{\text{MEY}}$$

$$(1.29)$$

$$C = 0: \qquad X^* = \frac{K}{2}\left(1 - \frac{\delta}{r}\right) = X^{\#}$$

which illustrate the facts that cost effects alone ($\delta = 0$) move X^* to the right of X_{MSY} and discounting alone moves X^* to the left.

Many alternative functional forms for the growth function $G(X)$ have been proposed (e.g., see Shepherd 1982b). For such forms, the golden-rule equation (1.26) may not have an analytic solution, but numerical solution (using Newton's method) is always straightforward. The special cases of Eq. (1.29) remain valid in general, although of course the analytic expressions are no longer correct.

Antarctic Baleen Whales

A notorious instance of resource depletion was provided by the twentieth-century pelagic whaling industry. By the 1960s all the largest species—blue, fin, right, gray, and humpback whales—had been reduced to fractions of their former abundance. Atlantic right whales were depleted during the nineteenth-century, as were Pacific gray whales. Stocks of the latter species in the northeastern Pacific, however, have recovered markedly as the result of protection since 1947. Catches of the major species in this century are shown in Figure 1.9. In terms of both weight and value of the catches, whales were for many years among the most commercially important of marine resources.

Whale catches have been controlled since 1947 by the International Whaling Commission (IWC). This agency, like all international bodies, depends on the cooperation of its member states for the establishment of and adherence to its regulations. (Nonmember states are of course under no obligation to adhere to such regulations, and this continues to be a matter of concern to the IWC.)

Critics of the whaling industry have argued that various "inadequacies" of the IWC convention have prevented the Commission from carrying out its mandate to conserve whale stocks. As pointed out elsewhere, however (Clark 1973a,b, 1976a; Clark and Lamberson 1982), the whaling companies, no doubt fully aware of the slow growth potential of whale populations, may have had little incentive to agree to conservation-oriented regulations. Many of the whaling companies have in

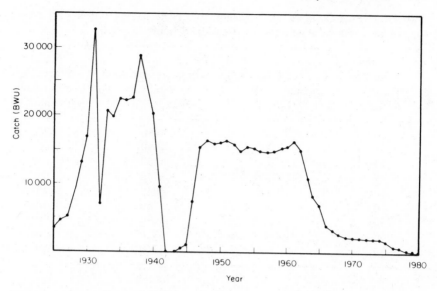

FIGURE 1.9 Total catch of Antarctic blue, fin, and sei whales, 1925–1980. (From Clark and Lamberson 1982, by permission of Butterworths, Guildford, U.K.)

fact diversified into other fisheries, following the demise of the whaling industry in the 1960s and 1970s.

As an illustration, let us apply the dynamic Gordon–Schaefer model to the Antarctic baleen whale fishery. We adopt the following parameter values, derived from IWC estimates:

$$r = 5\% \text{ per annum}$$
$$K = 400{,}000 \text{ blue whale units (BWU)}$$
$$q = 1.3 \times 10^{-5} \text{ per catcher day}$$

The BWU biomass unit, used by the IWC in setting quotas prior to 1963, was defined as follows: 1 blue whale = 1 BWU; 1 fin whale = $\frac{1}{2}$ BWU; 1 sei whale = $\frac{1}{6}$ BWU.

With the effort unit specified as one catcher day, we employ the following economic parameters:

$$\text{Price} = \$7500 \text{ per BWU}$$
$$\text{Variable effort cost} = \$5000 \text{ per catcher day}$$

But now we are faced with the problem of including *fixed costs*, that is, the capital costs of factory and catcher vessels, estimated at

$$\text{Capital cost} = \$2 \times 10^7 \text{ per standard factory fleet}$$

TABLE 1.2   Sole Owner's Optimal Biomass Levels, Yields, and Effort Levels for Antarctic Baleen Whale Stocks (Gordon–Schaefer Model)

Discount Rate δ (% per annum)	Optimal Biomass X^* (BWU)	Optimum Sustained Yield C^* (BWU per annum)	Optimal Effort E^* (catcher days per annum)
0	233,000	4860	1600
1	207,000	4990	1860
3	165,000	4850	2260
5	139,000	4540	2500
10	112,000	4040	2770
20	105,000	3880	2830
30	110,000	3990	2790
50	126,000	4320	2630
80	154,500	4740	2360

The Gordon–Schaefer model ignores fixed costs, which require a more sophisticated analysis—see Chapter 3. We here finesse the issue by charging interest and depreciation on capital as variable effort costs. One standard factory fleet exerts 2000 catcher days of effort per season. If γ denotes the annual depreciation rate, we therefore have

$$\text{Cost } c = \$[5000 + (\delta + \gamma) \cdot 10,000] \text{ per catcher day}$$

and we shall take

$$\gamma = 15\% \text{ per annum}$$

Note that effort cost c now depends on the discount rate δ, which we assume to be equal to the (real) rate of interest on capital. Consequently discounting no longer has the one-sided effect on X^* stated in the "fundamental principle."

Table 1.2 shows the optimal equilibrium whale stock level X^*, calculated from Eq. (1.28), as a function of the annual discount rate δ. The optimal sustained yield (OSY) $C^* = G(X^*)$ and the corresponding sustained effort $E^* = C^*/qX^*$ are also shown. Note that the discount rate has quite a pronounced effect on X^*, with "typical" (private firm) discount rates in the range 10–20% per annum leading to marked depletion of the whale stocks (MSY is at $K/2 = 200,000$ BWU). The effect on yields and effort levels is less pronounced—in fact, sustained optimal effort levels vary only from 0.8 to 1.4 "factory units" (of 2000 catcher days per year). The whaling industry actually constructed about

20 factory units after World War II, but this is not quite as "irrational" as it may seem—see Section 3.3.

Current stocks of baleen whales in the Antarctic are estimated to be about 75,000 BWU. This may be compared with the bionomic equilibrium biomass levels

$$\bar{X}_1 = 51,000 \text{ BWU} \quad \text{and} \quad \bar{X}_2 = 82,000 \text{ BWU}$$

corresponding to variable cost $c = \$5000$ per catcher day and total cost $c_{\text{total}} = \$8000$ per catcher day, respectively.

It seems therefore that either hypothesis—open-access bionomic equilibrium or profit maximization—can be supported by the data on Antarctic whaling. Because of the low growth potential and high commercial value of whales, the maximization of present value implies significant resource depletion, in agreement with the "fundamental principle" enunciated earlier (although the effect can be reversed in the case of unrealistically high rates of interest on capital—see Table 1.2).

The Social Rate of Discount

The general principle relating discounting to anticonservation, although easily captured in mathematical form, is much more than a mathematical theorem. It is a special case of the general economic principle that the demand for capital assets is inversely related to the interest rate. The importance of this principle to resource conservation has long been known and is discussed in the standard references (Ciriacy-Wantrup 1968, Clark 1976a, Dasgupta and Heal 1979, Herfindahl and Kneese 1974). Some authors have argued that considerations of intergenerational equity require the use of *zero* discount rate in public conservation decisions (Page 1976). Others have recommended 2–3% per annum as appropriate values for the social rate of discount; such rates reflect the real long-term rate of interest on secure investments such as government bonds (Feldstein 1964). *Private* real discount rates have been estimated at 10% per annum for large firms to as high as 39% p.a. for individuals (based on data on consumer behavior—see Hausman 1979).

In the case of fishery management, conscious use of any particular discounting policy is seldom if ever made. *Implicitly*, however, the perennial argument between fisheries biologists and the fishing industry—the former usually attempting to reduce catches and conserve fish stocks, the latter trying to increase current catch levels—is basically a disagreement over time discounting. The biologist–manager tends to adopt a long-term point of view, and hence a low discount rate, while the fisherman is much more concerned with his day-to-day survival, and hence implicitly uses a much higher discount rate. The uncertainties continually confronted by the fisherman only increase his implicit discount rate (Chapter 6). These uncertainties are greatly increased by government fisheries policy that permits and even encourages or ensures vast overcapacity.

Economic analyses of fishery management policy that suppress the intertemporal dimension may be highly misleading.

Other Social Considerations

The question might be raised whether maximizing economic benefits, discounted or not, is a proper objective for governments to adopt in fishery management. In a food-scarce world, shouldn't fisheries be managed for maximum yield? Yes—if food is the only scarce resource, but not necessarily if other, equally scarce resources are expended in catching fish. Shouldn't employment be the primary consideration? Productive employment, yes—but surely not employment that is counterproductive.

Should protection of the resource itself not be the first priority? Yes—provided resource protection does not imply human misery, as may happen if harvesting is prohibited entirely.

All these single-minded objectives—and others—have been put forth at one time or another as the "correct" view of fishery management. In truth, they conflict with one another. Economic analysis, combined with biological study, has the potential for assessing the costs and benefits of management alternatives and reaching an acceptable compromise. (See Chapter 4 for further discussion.)

1.3   An Example

In this section we describe briefly a bioeconomic model of a specific fishery, the Australian prawn fishery located in the Gulf of Carpentaria (Clark and Kirkwood 1979). The biological characteristics of tropical prawns render models of the Schaefer type inappropriate, but the bioeconomic principles discussed in this chapter are seen to persist in a different model. These simple principles are in fact quite robust. Thus although the Schaefer model may be biologically "unrealistic," its (cautious) use as a basis for general bioeconomic analysis seems justified. The mathematical simplicity of the Schaefer model adds to its attractiveness.

Tropical penaeid prawns typically have approximately a one-year life span. Recruitment is often highly variable and is not significantly related to spawning stock size. Individual prawns exhibit extremely rapid rates of growth and suffer high mortality rates. The following model (Beverton and Holt 1957) reflects these characteristics:

$$\frac{dN_t}{dt} = -(M + F_t)N_t \qquad (0 \le t \le T) \tag{1.30}$$

$$N_0 = R \tag{1.31}$$

$$w_t = w_\infty(1 - e^{-k(t-t_0)})^3 \tag{1.32}$$

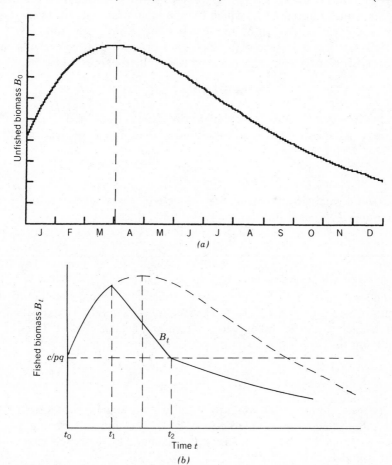

FIGURE 1.10 The Gulf of Carpentaria model: (a) unfished biomass; (b) fished biomass—fishing occurs for $t_1 \le t \le t_2$.

where N_t denotes the number of prawns at time t in a given year and $R = N_0$ denotes recruitment. M and $F_t = qE_t$ represent natural and fishing mortality, respectively, and T is the life span ($T \leq 1$ year). Recruitment at the beginning of the year, R, in actuality is a random variable, but we here treat it as a deterministic constant (see Chapter 6). The function w_i, called the von Bertalanffy growth curve, involves three parameters, w_∞, k, and t_0, which can be estimated from catch data.

Let $B_0(t)$ denote the unfished *natural biomass* of the prawn stock:

$$B_0(t) = N_t w_t = R w_\infty \, e^{-Mt}(1 - e^{-k(t-t_0)})^3 \qquad (1.33)$$

This function is illustrated in Figure 1.10(a). The biomass of the stock attains a maximum at $t = \bar{t}$, which for the main prawn stocks in the Gulf of Carpentaria occurs during the first week of April, approximately.

Next let p and c denote the ex-vessel price of prawns and the unit effort cost, respectively. Net revenue flow is then expressed as

$$\pi_t = (pqB_t - c)E_t \qquad (1.34)$$

where $B_t = N_t w_t$, with N_t specified by Eqs. (1.30) and (1.31). The industry net seasonal revenue (discounting over 1 year being ignored) is

$$\int_0^T \pi_t \, dt \qquad (1.35)$$

The sole-owner profit maximizing objective, for a given season, is to determine E_t so as to maximize (1.35), subject to

$$0 \leq E_t \leq E_{max} \qquad (1.36)$$

where E_{max} denotes maximum effort capacity (as determined by fleet size). Later we also determine the optimal capacity.

The solution to this optimization problem is simple and intuitively almost obvious [Fig. 1.10(b)—see Clark and Kirkwood 1979]. Fishing, at full capacity, should take place during a "fishing season" $t_1 \leq t \leq t_2$ enclosing the date $\bar{t}$ of maximum prawn biomass:

$$E_t = \begin{cases} E_{max} & \text{for } t_1 \leq t \leq t_2 \\ 0 & \text{otherwise} \end{cases} \qquad (1.37)$$

The season closes ($t = t_2$) when the fishery is no longer profitable, that is, when the fished biomass $B_f(t_2)$ equals c/pq. The optimal opening date t_1 (which depends on E_{max}) can easily be calculated by numerical simulation of the model.

The foregoing description in fact closely resembles the actual Gulf fishery on the important seasonal stocks of banana prawns (*Penaeus*

merguiensis). The optimal opening date t_1 is calculated from the above model. Fishing is prohibited prior to the opening date, and vessels continue fishing until catches decline to a nonprofitable level.

In the absence of any opening restriction, fishermen would be forced competitively to start fishing as soon as fishing became profitable—that is, at $t = t_0$ in Figure 1.10*b*. The average size of prawns in the catch would be considerably reduced,† and fishermen's revenue would decline correspondingly.

Optimal Fleet Size

Next let $V(E_{max})$ denote the total optimized seasonal net revenue, as given by Eq. (1.35). It is clear that $V(E_{max})$ is an increasing function of the fleet capacity E_{max} and approaches an asymptotic value as $E_{max} \to \infty$ (see Fig. 1.11).‡ Let c_f denote annual fixed costs (annual interest plus depreciation) per vessel. Total fixed costs are then $c_f \cdot E_{max}$. The (private

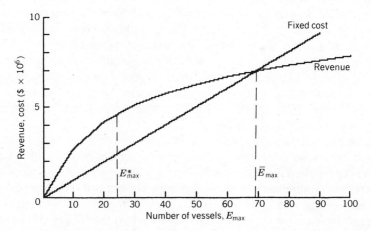

FIGURE 1.11 Seasonal net revenue and fixed cost in the Gulf of Carpentaria prawn fishery. Net fishery revenue is maximized at $E^*_{max} = 25$ vessels; open-access equilibrium occurs at $\bar{E}_{max} = 69$ vessels.

†Higher prices are paid for large prawns, and the model is easily adapted to account for this by letting $p = p_t$ be a function of time t.
‡It can be shown that $V(\infty) = \int_{c/pq}^{B_0(t)} (p - c/qB) \, dB$.

or social) optimum fleet size is that which

$$\text{Maximizes } [V(E_{\max}) - c_f E_{\max}] \tag{1.38}$$

(See Fig. 1.11.) Under open-access conditions, on the other hand, the number of vessels will reach equilibrium at $\bar{E}_{\max}$ where $V(\bar{E}_{\max}) = c_f \bar{E}_{\max}$.

Using bioeconomic data from the prawn fishery up to 1976, Clark and Kirkwood estimated the optimal fleet size as approximately 25 freezer trawlers and the bionomic equilibrium at 69 freezer trawlers. In actual fact, two distinct types of vessels, freezer trawlers and brine boats, participate in the fishery, and a more sophisticated model leads to the prediction of 50 freezer trawlers plus 90 brine boats at bionomic equilibrium (Clark and Kirkwood 1979, p. 1310). In 1976 there were 60 freezer trawlers and 85 brine boats.†

The purpose of introducing the Australian prawn model has been to indicate that (1) bioeconomic models differing considerably from the generic Schaefer model may be more appropriate in actual case studies, depending on particular biological and economic features of the fishery in question; and (2) the main *economic* predictions of the dynamic Gordon–Schaefer model pertain also to such alternative bioeconomic models. Biological overfishing, in the sense of stock depletion, does not occur, since the present model assumes that annual recruitment is independent of the previous year's escapement. *Economic* overfishing arises in two ways. First, under unregulated, open-access conditions, fishermen will catch prawns of suboptimal, immature size, thereby dissipating potential economic yield. This *observed* phenomenon is overcome by prohibiting fishing until the prawns have grown to an appropriate size. (The capture of suboptimally sized fish is sometimes referred to as *growth overfishing*.)

However, regulating the opening date of the fishery does not seem to prevent economic "overfishing," in the sense that if access to the fishery remains open, then fishing *capacity* (number of vessels) will increase until an economic equilibrium is achieved with costs equal to revenues. It is both interesting and significant that the first regulation (opening date) serves to *increase* the degree of overcapacity that the fishery can support. This prediction is not peculiar to the present model; rather it represents a general bioeconomic principle. We will be studying the question of overcapacity in some detail in Chapter 3.

---

†By 1980 nearly 300 (!) prawn vessels had received licenses to participate in the northern prawn fishery, and about 250 vessels did participate. A detailed economic survey appears in Ryan (1981).

1.4 Summary and Perspective

In this introductory chapter we have identified two leading and inter-related characteristics of the "open-access" fishery—depletion of fish stocks and overexpansion of the fishing industry. These phenomena can lead to significant economic losses, especially in valuable, highly productive fisheries.

A dynamic analysis indicates that successful remedies may be difficult to implement. Stock rehabilitation programs are painful and may encounter resistance from an established fishing industry. On the other hand, stocks successfully managed at high levels of biological productivity (e.g., MSY) will provide high catch rates, which may engender further and unnecessary expansion of fishing capacity. In the end, control over the fishery may be lost. Although these predictions are based here on a very simplistic model, there is every reason to expect that they are robust. Whether this is the case may be judged as the book progresses.

We have not yet, however, addressed the practical problems of fishery management in any depth. Before such a study can be undertaken, several questions of practical importance must be investigated. What is ultimately needed (and has hardly ever been attempted, at least until very recently) is a *predictive* theory of fishery regulation; for without such a theory the response of the fishing industry to imposed regulations will probably come largely as a surprise.

In order to predict, or understand, the response of the fishery to regulation, we need fairly explicit models of fishermen's behavior (not to mention the behavior of processors, marketers, regulators, etc.). Ideally, such models should also deal explicitly with *uncertainty*, which everyone recognizes as the basic reality in fishing.

The subsequent chapters of this book constitute an attempt to move toward a predictive theory of fishery management (Clark 1980a). No claim for completeness or even "maturity" of the theory can be made—fisheries are simply too complex ever to be finally understood.

Any theory must be field-tested, of course, but we are always forced to use theories which may ultimately turn out to be "wrong."† Unfortunately, most economically oriented fishery management techniques have received very little field testing to date. The situation appears to be changing rapidly, however, as coastal states are now faced with the management of major fishery resources that have recently come under their sole jurisdiction as the result of the new 200-mile zones. It will be

---

†According to Karl Popper, "correct" theories do not exist—or if they do, we have no way of recognizing them (Popper 1979).

unfortunate, indeed, if only the old tried-and-false methods of fishery management are brought to bear on these extremely valuable resources.

Appendix

We solve here the optimization problem for the full dynamic Gordon–Schaefer model, using the same elementary approach† (integration by parts) employed in Section 1.2 for a simple case. The problem is

$$\text{Maximize}_{\{E_t\}} \int_0^\infty e^{-\delta t}(pqX_t - c)E_t\, dt \tag{1.39}$$

subject to the conditions

$$\frac{dX_t}{dt} = G(X_t) - qE_tX_t \tag{1.40}$$

$$0 \le E_t \le E_{\max} \tag{1.41}$$

$$X_t \ge 0 \tag{1.42}$$

Also, the initial biomass X_0 is assumed known.

First we solve for E_t in (1.40) and substitute into the objective integral (1.39), which then becomes

$$\int_0^\infty e^{-\delta t}[p - c(X_t)] \cdot \left[G(X_t) - \frac{dX_t}{dt} \right] dt \tag{1.43}$$

where $c(X) = c/qX$. Define the function $Z(X)$ as follows:

$$Z(X) = \int_{\bar X}^X [p - c(u)]\, du \tag{1.44}$$

where $\bar X = c/pq$. Note that for $Z_t = Z(X_t)$ we have

$$\frac{dZ_t}{dt} = [p - c(X_t)]\frac{dX_t}{dt}$$

Integration by parts therefore yields

$$\int_0^\infty e^{-\delta t}[p - c(X_t)]\frac{dX_t}{dt}\, dt = \int_0^\infty e^{-\delta t}\frac{dZ_t}{dt}\, dt = e^{-\delta t}Z_t \Big|_0^\infty + \delta \int_0^\infty e^{-\delta t}Z_t\, dt$$

$$= Z(X_0) + \delta \int_0^\infty e^{-\delta t}Z_t\, dt$$

---

†This problem also yields to the famous maximum principle of optimal control theory, which becomes necessary if the model is altered in any essential way—see Chapter 3.

Since the fixed term $Z(X_0)$ can be ignored, our integral (1.43) now becomes

$$\int_0^\infty e^{-\delta t} V(X_t)\, dt \tag{1.45}$$

with

$$V(X_t) = [p - c(X_t)] \cdot [G(X_t) - \delta X_t] \tag{1.46}$$

But obviously (1.45) is maximized if $X_t = X^*$, where X^* maximizes this expression $V(X_t)$, and the biomass X_t should be adjusted from X_0 to X^* (by fishing at $E_t = 0$ or E_{max} as required) as rapidly as possible.

A straightforward calculation shows that the necessary condition $V'(X^*) = 0$ can be expressed in the form

$$G'(X^*) - \frac{c'(X^*)G(X^*)}{p - c(X^*)} = \delta \tag{1.47}$$

which is the desired result.

[It should be noted here that the above argument, to the effect that X_t should be adjusted by fishing to X^*, tacitly assumes that the function $V(X)$ is "monodromic," i.e., that $V(X)$ has a *unique* local maximum on $\bar{X} \le X \le K$. It also assumes that E_{max} is sufficiently large that X_t *can* be adjusted to and kept at X^*. Strange—and quite interesting—things happen if either of these assumptions is violated; see Spence and Starrett (1975) and Majumdar and Mitra (1983).]

Finally we show that X^* is a decreasing function of δ, at least for the Schaefer model per se, in which

$$G(X) = rX\left(1 - \frac{X}{K}\right) \quad \text{and} \quad c(X) = \frac{c}{qX}$$

In this case the left side of (1.47) becomes

$$r\left(1 - \frac{2X}{K}\right) + \frac{cr(1 - X/K)}{pqX - c}$$

This is obviously a decreasing function of X for $X > \bar{X} = c/pq$, and its value approaches $+\infty$ as $X \to \bar{X}$ from above. It therefore follows from (1.47) that X^* is a decreasing function of δ and also that $X^* \to \bar{X}$ as $\delta \to +\infty$.

2 Search and Capture

In the models presented in Chapter 1, a very simple relationship was adopted between input, in the form of "fishing effort," and output, in terms of the rate of catch of fish. The fundamental assumption incorporated in the Schaefer model [see Eqs. (1.3)–(1.7)] was that fishing mortality F was directly proportional to fishing effort E, or more precisely that fishing effort could be quantified in some fashion so that the assumed relation was valid. In this chapter, the relationship between catch and effort will be modelled and analyzed in greater detail.

The first matter requiring attention is a clarification of the basic concept of fishing effort. Frequently the meaning of this term seems to be considered self-evident. Economic studies of fisheries, for example, seldom attempt to define effort operationally, referring to it simply as an index of capital and labor input. Yet biologists are far from reaching agreement on a uniform definition of fishing effort. For example, several authors concerned with the management of fluctuating fish stocks have recommended the regulation of effort as superior, for various reasons, to the regulation of catch (Beddington and May 1977, Reed 1981, Sissenwine and Kirkley 1982). Upon examination, however, it turns out to be fishing *mortality*, rather than effort, that is being referred to in these recommendations.

Obviously if fishing mortality and fishing effort are indeed directly proportional, then control of one will be equivalent to control of the other. Unfortunately, however, there is little reason to be confident that fishing effort and mortality are closely related in many fisheries. For certain pelagic species, such as herrings and anchovies, it is now generally agreed in fact that the relationship is likely to be very weak. Because of the schooling behavior of these species, catch per unit effort tends to remain high at quite low population levels, and this is now seen as an important factor contributing to the long history of collapse of pelagic fisheries (Saville 1980). The second main topic of this chapter, therefore, is an investigation of the overall relationship between catch and effort, culminating in the important concept of the *concentration profile*.

From the viewpoint of individual fishing vessels, catches per trip—and likewise annual catches—fluctuate to such a degree as to render any deterministic model of fishing highly unrealistic. Since any form of fishery regulation operates through the activities of individual vessels, a preditive theory of fishery management must of necessity include fluctuations in catch rates. The modelling of the capture of fish (and the search for fish) as a random or stochastic process is therefore a third leading concern of this chapter. Our analysis indicates that searching for high fish concentrations is a major consideration for the fisherman and thus emphasizes the importance of concentration profiles.

2.1 Fishing Effort

In this section we present a series of definitions pertaining to the concept of fishing effort. These definitions represent an attempt to abstract and uniformize discussions of fishing effort in the literature (cf. Gulland 1964a, Pope 1975, Rothschild 1977). Initially the definitions should be thought of as applying to the case where fishing gear is drawn through the water, as in trawling for fish; other fishing techniques will be discussed later.

Our basic definitions are not intended necessarily to be directly operational. Nevertheless, they should help considerably to clarify the difficulties involved in actually measuring fishing effort and relating it to fishing mortality.

Definition 1. *e = fishing effort*: volume of seawater screened by the fishing gear per unit time. Unit: m³/h. (See Treschev 1975.)

A given vessel–gear combination, operating at the vessel's standard cruising speed, thus exerts a specific level of fishing effort during fishing operations. Any particular such combination can be taken as a *standardized vessel–gear unit* or, more simply, a *standardized fishing unit* (SFU).

Definition 2. *E = nominal fishing effort*: number of standardized vessel-gear units actively fishing at a given time. Unit: SFU.

Depending on the unit of standardization, we therefore have

$$e = aE \qquad (2.1)$$

where *a* = constant represents the volume of seawater screened per hour by 1 SFU.

Fishing effort can be aggregated over any given area of the ocean, *A*, and over any given time interval *T*, with vessel *i* exerting nominal effort E_i for τ_i hours' duration. Then the *aggregate fishing effort* $\bar{e}$ is defined by

$$\bar{e} = a\bar{E} = a \sum_{i=1}^{N} \tau_i E_i \qquad (2.2)$$

Note that $\bar{e}$ has units m^3 and represents the total volume of seawater screened by the fleet within the given time period. [This formulation presupposes that the *i*th vessel, at any given time, is either not fishing at all or fishing at nominal effort E_i. If E_i can be varied continuously, the expression $\tau_i E_i$ should be replaced by $\int_0^T E_i(t)\, dt$.]

Definition 3. ϵ = *selectivity factor*: proportion of fish in the volume *V* captured by the fishing gear. Unit: dimensionless ($0 \le \epsilon \le 1$).

Definition 4. ρ = *density* (or *concentration*) of fish within *V.* Unit: kg/m³.

If the *structure* of the catch is of interest (e.g., age or size composition, species composition, etc.), then both ϵ and ρ will be appropriate vector quantities (or even matrix quantities). For example, ρ_{ij} could represent the density of age-class *i* fish of species *j*.

It is now a logical consequence of these definitions that, for an individual fishing unit, the instantaneous rate of catch *C* is given by

$$C = \epsilon a E \rho \quad \text{kg/hr} \qquad (2.3)$$

where ρ denotes the density of fish (of a particular type) encountered by the gear at the time in question.

Equation (2.3) asserts that, for the individual fishing unit, *catch per unit effort is proportional to the stock density* encountered by the gear. [If gear is fished to the point of saturation—for example, by blocking of net interstices—then the selectivity factor ϵ in Eq. (2.3) may vary over time.] In this highly restricted sense, the Schaefer hypothesis becomes a tautology. Difficulties arise when one wishes to aggregate catch data over time and space—as must be done in practice, of course. In the unlikely event that ρ *is everywhere constant*, aggregation is achieved simply by summing (2.3) over all fishing units, to obtain a total catch rate of

$$C_{\text{total}} = q E_{\text{total}} X \quad \text{kg/hr} \tag{2.4}$$

where $X =$ total stock size (kg) and $q = \epsilon a / V_{\text{tot}}$, with $V_{\text{tot}} =$ volume of seawater occupied by fish, and $E_{\text{total}} = \sum_i E_i$. Here the *catchability coefficient q* has dimensions $(\text{SFU hr})^{-1}$.

In general, when N vessels operate, total catch rate is given by

$$C_{\text{total}} = a \sum_{i=1}^{N} \epsilon_i E_i \rho_i \tag{2.5}$$

where ρ_i denotes the stock density encountered by the ith vessel, and ϵ_i is its selectivity factor. [In the case that the catch is structured (by size, species, etc.), Eq. (2.5) becomes a system of equations, one for each component in the structure.] Note that even if all vessels have the same selectivity factor $\epsilon_i = \epsilon$, the sum in (2.5) cannot be reduced to *any* functional relationship of the form

$$C_{\text{total}} = F(X, E_{\text{total}}) \tag{2.6}$$

(where $E_{\text{total}} = \sum_i E_i$) *unless ρ_i is also the same for all vessels*. Thus the Schaefer hypothesis $C_{\text{total}} = q X E_{\text{total}}$ involves, inter alia, a tacit assumption of equal concentration ρ throughout the fishing area.

Definition 5. $f =$ *instantaneous fishing mortality*: proportion of the total fish stock biomass X removed by fishing, per unit time. Unit: hr^{-1}.

Thus by definition,

$$f = \frac{C_{\text{total}}}{X}.$$

From the foregoing discussion it follows that in general there is *no* relationship between fishing mortality f and total effort E_{total}.

One approach to resolving this difficulty of aggregating catch and effort data involves a "localization" of the data. That is to say, the whole fishing area A occupied by the stock is subdivided into smaller areas A_j, in such a way that the stock density ρ_j in each subarea can be approximated by a constant (Gulland 1956, Calkins 1961). If V_j denotes the volume of water encompassed by A_j, then as in Eq. (2.4) we have

$$\frac{V_j C_j}{E_j} = \epsilon a X_j$$

and by summation

$$\sum_j V_j \left(\frac{C_j}{E_j}\right) = \epsilon a X \qquad (2.7)$$

In other words, the weighted sum $\sum_j V_j(C_j/E_j)$ becomes an index of

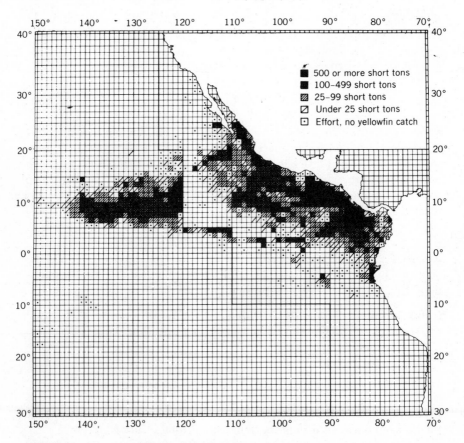

FIGURE 2.1 Catches of yellowfin tuna in the eastern Pacific Ocean in 1974 by 1° areas. (From Inter-American Tropical Tuna Commission 1975.)

relative abundance. (Two technical problems with this approach are, first, that the A_j must be large enough for the catch samples to be statistically significant and, second, that the expression C_j/E_j is undefined if no fishing takes place within A_j. A Bayesian approach to the latter difficulty is discussed in Section 2.4.)

The foregoing method is regularly used to obtain indices of stock abundance for yellowfin and skipjack tuna in the eastern tropical Pacific Ocean (Inter-American Tropical Tuna Commission 1975). This fishery covers an area of over 5 million km², and tuna concentrations are obviously far from uniform over this area (see Fig. 2.1).

An alternative approach to the aggregation problem will be described in Section 2.2.

Other Fishing Techniques

The variety of fishing techniques in use is so great that no single definition of effort can be expected to apply to all situations. But if fishing effort is to bear any relationship to stock density, it is clear that the definition must be based on the rate at which the fishing area is screened for fish. In many fisheries a large proportion of time at sea is spent in *searching* for schools of fish, which are then captured in a separate operation. In such cases, fishing effort is best interpreted as *searching effort*, defined as the volume of water (or area of surface or bottom, where appropriate) searched per unit time. Various complications, such as onset of darkness, set time, probability of detection and capture, effects of weather conditions, and size distribution of schools can also be taken into account (Neyman 1949, Pella and Tomlinson 1969, Mangel 1981).

Certain types of gear, such as traps and long lines, are left in place for some time before being lifted and stripped of fish. The physical dimensions and other characteristics of such devices, as well as the length of time between placement and lifting, will obviously influence their catching power. However, fishing effort can still be defined as the volume of seawater from which the fish are removed per unit time.

In principle, then, it seems possible to adapt our original definition of fishing effort (i.e., rate of screening of the fishing ground) to most fishing techniques. If so, we obtain by definition the general rule:

$$\text{Catch rate per unit effort} \propto \text{stock abundance}$$

but only on a *local* scale, that is, for each separate fishing unit for each point in time. Unless stock density is the same for all fishing units, this relationship cannot be extended to the aggregate case, as noted above. This difficulty is addressed in the following section.

2.2   Concentration Profiles

It is of course unlikely that all vessels of a given fleet will encounter the same density of fish stocks simultaneously. Nevertheless, economic incentives do exist which should bring about some degree of convergence. Namely, all vessels have an incentive to *maximize* their catch rates—subject, of course, to considerations of costs involved in searching for the maximum concentration of fish and in shifting the vessel from one location to another. Of course, fishermen's skills vary, and some may be more successful than others.

In this section we investigate the consequences of assuming that fishermen always exploit the highest existing concentrations of fish (Clark 1982), recognizing that in practice this would be at best an approximation. Under this condition we have $\rho_i \equiv \rho$ for all active fishing units, and therefore Eq. (2.5) becomes

$a = \dfrac{M^3}{SFU} \cdot \dfrac{K}{M^3}$

$$C = a\rho \sum_{i=1}^{N} \epsilon_i E_i \tag{2.8}$$

(where we now write C for C_{total} = total catch rate).

The expression $\sum_{1}^{N} \epsilon_i E_i$ is the *effective total nominal effort* of the active fleet of N vessels (unit: SFU). However, to simplify notation, we also assume that $\epsilon_i \equiv \epsilon$, so that Eq. (2.8) can be written as:

$$C = \theta E \rho \tag{2.9}$$

where C = total catch rate (kg/hr)
$\quad\quad E$ = total nominal effort (SFU)
$\quad\quad \rho$ = maximum stock density (kg/m^3)
$\quad\quad \theta = a\epsilon$ = effort scale factor (m^3/SFU hr).

Equation (2.9) will be basic to the remainder of the discussion in this section.

Following the customary terminology, we formulate

Definition 6. Under the assumptions leading to Eq. (2.9) we define the (nominal) *catchability coefficient* q as

$$q = \frac{\theta \rho}{X} \quad \text{(per SFU hr).} \tag{2.10}$$

Consequently we obtain the familiar equation

$$C = qEX \tag{2.11}$$

but with a coefficient q which is in general a function of the current stock size X.

We now consider the development of a fishery, as effort is applied and the stock biomass X is reduced. We continue to assume that the fishermen always exploit the highest concentrations first. The question is, what relationship, if any, exists between this concentration ρ and the current stock level X? (Here we ignore any population structure and treat X as a single, lumped variable. Interesting complications arise when structure is introduced—see Chapter 5.)

To focus on this question, we first imagine a completely immobilized resource stock X (such as a mineral ore), spread over space with variable density ρ. Let $f(\rho)\, d\rho$ denote the total amount (kg) of the resource existing at concentrations (densities) between ρ and $\rho + d\rho$. (An appropriate spatial scale is considered, so that the ultimate granular structure of the resource is not relevant.)

A typical assumption in the mining industry, for example, is that $f(\rho)$ is a decreasing function of ρ [Fig. 2.2(a)]—that is, the lower the concentration the greater the reserves. Let $X(\rho)$ denote the total reserves existing at concentrations $\leq \rho$:

$$X(\rho) = \int_0^\rho f(\rho)\, d\rho \tag{2.12}$$

Then, for $f(\rho)$ decreasing, we have $X'(\rho) = f(\rho) > 0$ and $X''(\rho) = f'(\rho) < 0$; that is, $X(\rho)$ is increasing and concave downwards [Fig. 2.2(b)].

We now invoke our hypothesis that highest concentrations are

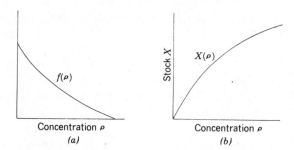

FIGURE 2.2 Construction of (inverse) type I concentration profile $X(\rho)$. Here $X(\rho)$ is the integral of $f(\rho)$—see text.

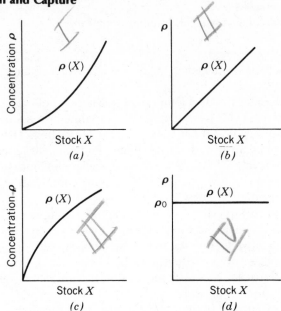

FIGURE 2.3 Concentration profiles $\rho(X)$ of four types. (a) Type I (convex); (b) type II (linear); (c) type III (concave); (d) type IV (constant concentration). (Other types, involving S-shaped profiles, could also be considered.)

exploited first. The exploitation history will then follow the curve $X = X(\rho)$ or, perhaps more accurately, the inverse curve $\rho = \rho(X)$, in the sense that when the stock level is at X the concentration being exploited will be the marginal (maximum) concentration $\rho(X)$. We refer to this curve $\rho = \rho(X)$ as the (*marginal*) *concentration profile*. Thus $\rho(X)$ represents the maximum density of the resource when the stock level is X; as X is depleted this concentration decreases, by our assumption. Hence $\rho'(X) \geq 0$. The case shown in Figure 2.2(b), and redrawn in inverse position in Figure 2.3(a), has $\rho''(X) > 0$ for all X; this is called a *type I* (*convex*) concentration profile.

Other types of profiles may exist—especially for "diffusive" resources such as fish stocks (see below). Since $f(\rho) \geq 0$, the concentration profile is increasing in X (i.e., the marginal concentration necessarily declines as X is reduced). The *type II* (*linear*) profile of Figure 2.3(b) arises when $f(\rho) \equiv f_0 = $ constant—that is, when the amount of the resource at each concentration ρ is the same. (This type, which may seen unrealistic, is important in the fishery case.)

The direct opposite of type II is the case in which the entire resource stock exists at a fixed, constant concentration $\rho = \rho_0$ [Fig. 2.3(d)]. We refer to this extreme case as *type IV* (*constant concentration*).†

A less extreme case, *type III* (*concave*), arises when the amount of the resource at concentration ρ, that is, $f(\rho)$, is a monotonically increasing function of ρ [Fig. 2.3(c)]. Even more complex (convex–concave) concentration profiles can be constructed, but we leave this to the imagination of the reader.

Suppose now that the resource stock is exploited at a nominal effort level E (possibly varying over time—this is irrelevant), where as always "effort" is the rate at which the sea water is screened for the resource. We then have, as in Eq. (2.9),

$$C = \theta E \rho(X) \tag{2.13}$$

where C designates the rate of recovery of the resource stock. This can also be written as [cf. Eq. (2.11)]

$$C = q(X)EX \tag{2.14}$$

where the catchability coefficient $q(X) = \theta\rho(X)/X$ is now specifically a function of the total stock size X. In other words,

$$\frac{C}{E} \propto \rho = \rho(X) \tag{2.15}$$

that is, recovery per unit effort is proportional to the marginal stock density. Thus the concentration profiles of Figure 2.3 can also be interpreted as curves of C/E vs. stock abundance X.

The hypothesis that exploitation of the resource follows the (marginal) concentration profile therefore allows us to *aggregate* "catch" (recovery rate) and effort data to obtain an index of stock abundance. Except for type II profiles, however, the index is nonlinear.

We can now see what sort of bias would be introduced by assuming, as in the Schaefer model, that $C/E \propto X$. For type I profiles, the remaining stock X would be progressively *underestimated* by this procedure, whereas for type III or IV profiles it would be *overestimated*. (Indeed, for type IV profiles the depletion of X would now show up at all until the resource had been completely exhausted!) Only for the apparently unlikely case of linear Type II profiles would the estimate of X be unbiased.

The possible bias inherent in the Schaefer model can only be removed, even in a qualitative sense, if some a priori idea of the concentration

---

†For type IV, the function $f(\rho)$ is a Dirac delta function, $f(\rho) = X_0\delta(\rho - \rho_0)$, where X_0 = initial reserves.

profile is available. What is to be expected in the case of fishery resources?

Diffusive Resource Stocks

In order to move a little closer to the spatial properties of fish populations, we next consider the possibility of diffusion of stocks from areas of high concentration to areas of lower concentration. An ideal case arises when the resource constitutes a single pool and diffusion is so rapid, relative to the rate of depletion, that the stock density always remains uniform throughout the pool. Then we have simply $\rho = $ constant $\times X$, that is, a type II linear profile. This assumption of rapid diffusion to a uniform density thus underlies the traditional Schaefer model $C = qEX$ ($q = $ constant).

More generally, suppose there are n separate, disconnected pools, each with the above rapid-diffusion property and concentrations

$$\rho_i(t) = a_i X_i(t) \qquad (a_i = \text{constant})$$

Suppose initially that $\rho_1(0) \geq \rho_2(0) \geq \cdots \geq \rho_n(0)$. Then the first pool will be exploited exclusively until $\rho_1(t) = \rho_2(0)$, after which pools 1 and 2 will be exploited simultaneously until $\rho_1(t) = \rho_2(t) = \rho_3(0)$, and so on. For $X = \sum_1^n X_i$, it is easy to see that $\rho(X)$ is piecewise linear, of type I (see Fig. 2.4), the slope of the kth segment from the right being $(\sum_1^n a_i^{-1})^{-1}$.

Next suppose that the n pools are interconnected and that the rate of diffusion between pools is determined by their relative concentrations. A model of this situation for $n = 2$ pools is

$$\left. \begin{aligned} \frac{dX_1}{dt} &= k(X_2 - mX_1) - Q_1 X_1 \\[2mm] \frac{dX_2}{dt} &= -k(X_2 - mX_1) - Q_2 X_2 \end{aligned} \right\} \qquad (2.16)$$

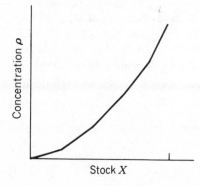

FIGURE 2.4 Concentration profile for n unconnected resource pools.

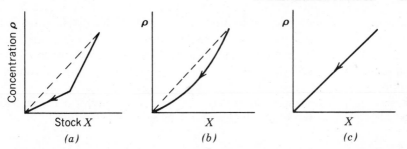

FIGURE 2.5 Concentration profiles for two connected pools, for $0 \leq k \leq \infty$ (a) $k = 0$; (b) $0 < k < \infty$; (c) $k = +\infty$. As the diffusion coefficient k approaches ∞, the type I profile converges to type II.

where k, m are positive constants and Q_1, Q_2 denote the exploitation rates ("fishing mortalities") of the two stocks. The analysis of this system of equations, under the additional marginal density hypothesis, is fairly straightforward, but we omit the details (see Clark 1982). The concentration profile is again of type I, with specific form depending on the diffusion constant k (see Fig. 2.5). For $k = 0$ there is no diffusion, and the result is the same as in Figure 2.4; on the other hand, as $k \to +\infty$, the two pools essentially become a single pool, and the concentration profile approaches the linear type II case for a single pool.

The following qualitative conclusions can be drawn from the analysis so far. First, a *sedentary* or slowly diffusing fish stock, if exploited according to our marginal concentration hypothesis, will tend to exhibit a type I (concave) concentration profile, with stocks of high density being the first to be depleted, and with catch per unit effort (CPUE) falling more rapidly than the stock X itself. (An exception arises if the density is everywhere the same, i.e., there is a single pool. In this case, the model assumption of infinite diffusion *within* the pool is unrealistic for the sedentary case. In fact, a type IV profile is appropriate to this case—see below. More complicated yet is the possibility of several large sedentary pools; the result is a profile of mixed type.)

Gulland (1964b) gives the following example. North Sea stocks of demersal fish (plaice, cod, and sole), largely unexploited during World War II, yielded high catch rates when fishing recommenced after the war. CPUE in these fisheries fell rapidly in the first two postwar years, but on the basis of cohort analysis,† Gulland demonstrates that the stocks

---

†Cohort analyses [and the related technique of virtual population analysis (VPA)] employs the historical record of catches and their age composition to reconstruct the initial size and subsequent history of each "cohort," or age class, to have passed through the fishery. Roughly speaking, the initial population is estimated as the sum of all catches from the cohort and fish dying from natural causes. The accuracy of the method depends largely on a complete catch history of the cohort as well as on the ability to estimate natural mortality. See Pope (1972).

themselves did not decline to the same degree. The discrepancy was attributed to the fact that dense pockets of fish were quickly located and fished out, after which the main stock, having a lower concentration, was exploited. These sedentary species therefore exhibited a type I concentration profile, at least during this phase of postwar adjustment. Ultimately the concentration must have become more uniform throughout the North Sea as the high-concentration areas were fished out.

Aggregative Resource Stocks

Diffusion implies that a substance flows outward from areas of high concentration. Many fish species, however, behave in the opposite fashion, continually forming into large, dense schools. In many cases, the schools also tend to congregate. The evolutionary advantages of schooling may include protection from predation, improved foraging (Clark and Mangel 1984), improved migration or spawning success, and so on (Breder 1967, Shaw 1970, Murphy 1980). Whatever the natural advantages, it is clear that schooling behavior is also extremely important to the fishermen.

As the total size of a schooling population is reduced by fishing, either the average size of schools or the total number of schools (or both) must diminish. In most cases, available data seem to indicate that the main reduction occurs in the number rather than in the size of schools. The effect of schooling behavior on concentration profiles depends upon the ease with which the remaining schools are located. For fast-swimming, predatory, pelagic species such as tuna, the spatial distribution of schools may retain roughly the same pattern, independent of total abundance. As demonstrated above, this would lead to a type I or type II concentration profile. [Another aggregation characteristic of tuna, partial schooling, is analyzed by Clark and Mangel (1979).]

An important group of pelagic schooling species is the clupeoids: anchovies, sardines, pilchards, herrings, and like species. These naturally abundant species exploit the primary production of the ocean and themselves constitute a major food source for fishes at higher trophic levels. They have supported many of the world's largest fisheries but have also exhibited a general tendency to collapse under heavy fishing pressure (Murphy 1977, Saville 1980)—see Table 1.1. In some cases, populations have not recovered, or recovered only very slowly, after fishing has ceased. As noted in Chapter 1, these pelagic species also appear to be subject to large-scale natural fluctuations of abundance, often correlated with major environmental events such as changes in upwelling pattern.

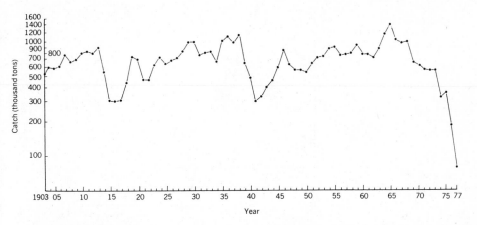

FIGURE 2.6 Total landings of North Sea herring, 1903–1977. (From Saville and Bailey 1980.)

Heavy fishing, however, almost certainly increases the potential for environmentally induced collapse.

The huge schools formed by clupeoid species are easily located by modern search technology, especially near spawning grounds, where they are often fished. The spacing of fish within schools is largely determined by the size of the fish, so that school density is more or less constant. Thus, whether fishing effort is equated to searching rate or screening rate of gear, CPUE is likely to remain fairly constant, regardless of population size. This result—a type IV concentration profile—is sometimes referred to as the Paloheimo–Dickie catch–effort relation (Paloheimo and Dickie 1964).

Empirical evidence in support of this conjecture is given, for North Sea herring stocks, by Pope (1980) and by Saville and Bailey (1980); see also Ulltang (1980). Catches of North Sea herring, 1903–1980, are shown in Figure 2.6. Using virtual population analysis, Pope (1980) has estimated fishing mortality for the period 1964–1974 and has compared fishing mortality with catch per unit effort for various gear types—see Figure 2.7. Ulltang (1980) presents evidence to indicate that, if a catchability-population relationship of the form

$$q = kX^{-b} \tag{2.17}$$

is assumed, a value of $b \approx 1$ gives the best fit to data for Norwegian spring herring. McCall (1976), on the other hand, estimated $b \approx 0.61$ for the California sardine fishery, 1937–1944.

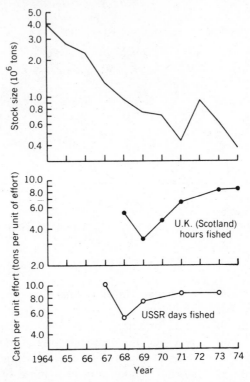

FIGURE 2.7   Index of abundance and carch per unit effort for two purse-seine fleets for North Sea herring, 1964–1974. (From Pope 1980.)

Note that, by Eqs. (2.13) and (2.14), the relationship (2.17) implies that

$$\rho(X) \propto X^{1-b} \tag{2.18}$$

This corresponds to a type I profile if $b < 0$, type II if $b = 0$, type III if $0 < b < 1$, and type IV if $b = 1$. The parameter b can be estimated (under the marginal concentration assumption) by combining catch–effort data with independent data on stock abundance, as obtained, for example, from cohort analysis.

Other Influences

Fish concentrations are obviously likely to change over time for reasons other than removal by fishing vessels. Some species, such as salmon and tuna, have traditional migration patterns which largely determine fishing strategy. Often, concentrations are highest on or near spawning grounds, in which case uncontrolled fishing may have particularly severe consequences for the survival of the fishery.

The assumption that fishermen always exploit the most concentrated stocks is obviously oversimplistic. For example, locating such concentrations may be costly, and fishermen may prefer the "bird in hand"—to mix metaphorical taxa. The question of search strategy will be discussed in Section 2.4.

Fish stocks may be severely affected by predation, so that changes in the concentration of different species may be interrelated. This difficult topic will be discussed in Chapter 5.

In many fisheries, the operation of fishing vessels is divided between searching and "handling time," that is, time spent capturing fish that have been located. This phenomenon, when accompanied by fluctuations in the catchability coefficient, can severely bias the relationship between CPUE and stock abundance in the direction of a type III profile (see Cooke 1984, who also discusses many other sources of bias).

2.3 Management Implications

The management of fisheries based on pelagic schooling species, such as the clupeoids, has proven to be particularly difficult. If the arguments of the previous section are valid, these fisheries can be expected to exhibit type III–IV concentration profiles. For such fisheries, CPUE remains constant, or diminishes only slightly, as the stock is fished down. Consequently the usual "signal" of stock depletion, namely declining catch rates, is absent, both to the managing authority and to the fishing industry itself.

In a detailed review of the southwest African pilchard fishery, Butterworth (1980) poses the question: Could the 1975–1978 collapse of this stock (a primary herbivore in the highly productive Benguela upwelling system) have been predicted on the basis of data available to the management authority? Butterworth considers four approaches to stock estimation: CPUE indices, virtual population analysis (VPA), modal length indices, and direct stock surveys. He concludes that none of these approaches could have predicted the collapse with any degree of certainty. "The use of CPUE as a biomass index is based on the assumption that [fishing mortality] F is linearly proportional to E .... In [a] pelagic context, with the stock concentrated in relatively few shoal groups in specific areas, the accuracy of such a model is open to doubt" (Butterworth 1980, p. 70). The VPA method is "increasingly less accurate for more recent years" (p. 70) and did not indicate the decline in recruitment until 1977–1978. The modal length (i.e., modal length of fish in the catch) rose from 1970 to 1976 and only began to reflect stock decline in 1977. Direct surveys (aerial estimates and egg surveys) were carried out only up to 1974 and are known to be rather unreliable, especially for highly clumped populations (cf. Swierzbinski 1981).

These conclusions are pessimistic—how can a fishery be managed at all if no method exists for predicting imminent collapse? Intuitively, such uncertainty would seem to demand a cautious approach to exploitation. This important question will be taken up in Chapter 6.

Bioeconomic Models

The existence of a concentration profile $\rho(X)$ other than type II (i.e., Schaefer) can have a profound influence on the predictions provided by the bioeconomic model discussed in Chapter 1. The model now becomes

$$\frac{dX_t}{dt} = G(X_t) - C_t \tag{2.19}$$

$$C_t = q(X_t)E_tX_t \tag{2.20}$$

where [see Eq. (2.11)]

$$q(X) = \frac{\theta\rho(X)}{X} \tag{2.21}$$

Let $c(X)$ again denote the cost of effort required to catch one unit of fish when the stock level is X. Then, assuming cost proportional to effort, we have

$$c(X) = \frac{c}{Xq(X)} = \frac{c}{\theta\rho(X)} \tag{2.22}$$

since $C_t \Delta t = 1$ implies cost $= cE_t \Delta t = c/X_t q(X_t)$. Gordon's net revenue flow expression then becomes

$$\pi_t = pC_t - cE_t$$
$$= [pq(X_t)X_t - c]E_t$$
$$= [p - c(X_t)]C_t \tag{2.23}$$

Bionomic equilibrium of the open-access fishery occurs at $X = \bar{X}$, where

$$c(\bar{X}) = p, \quad \text{that is,} \quad \rho(\bar{X}) = \frac{c}{\theta p} \tag{2.24}$$

unless $\rho(X) < c/\theta p$ for all feasible stock levels X, in which case bionomic equilibrium occurs at $X = K$ (the carrying capacity) and has $E = 0$. The density $\bar{\rho} = \rho(\bar{X})$ can be considered the "break-even" stock density, with the property that fishing is not worthwhile when $\rho < \bar{\rho}$.

The bionomic equilibrium prediction takes on an interesting form for the case of a type IV concentration profile, with $\rho(X) \equiv \rho_0 = $ constant. In this situation there exists a *threshold* price/cost ratio $(p/c)_{thr}$ given by

$$(p/c)_{thr} = \frac{1}{\theta\rho_0} \qquad (2.25)$$

and having the property that

$$X = \begin{cases} K & \text{if } \frac{p}{c} < \left(\frac{p}{c}\right)_{thr} \\ 0 & \text{if } \frac{p}{c} > \left(\frac{p}{c}\right)_{thr} \end{cases} \qquad (2.26)$$

Thus *a type IV fishery is either not viable at all, or else the unregulated fishery results in extinction of the resource population*! If the price/cost ratio increases over time, the fishery will develop suddenly and then rapidly be fished to extinction.

Obviously this prediction is a bit oversimplistic—no fish population is known to have been fished to actual extinction (probably no population has a strict type IV profile). Also, environmental changes appear to have been implicated to some extent in most recent recorded collapses. Nevertheless, the type IV scenario does come close to fitting the facts for quite a number of the pelagic fisheries listed in Table 1.1, several of which developed and collapsed rather rapidly in the 1960s and 1970s—the extreme example being perhaps the Peruvian anchoveta (Fig. 2.8). (The anchoveta collapse may also have been associated with environmental conditions—namely the recurring change in tropical currents referred to as El Niño—but that is not the issue here.)

In Chapter 1 we mentioned briefly the possibility of employing taxes or royalties on catch to force the open-access fishery to adopt a desired rate of exploitation. There are many practical problems with this approach, not least of which would be the determination of the appropriate tax schedule. Because of the threshold effect, control of a type IV stock would offer special difficulties. Too large a tax would shut down fishing activity entirely [see Eq. (2.26)], whereas a tax that was too small would have no effect at all. (A tax on effort E would have the same difficulty, because of the type IV relationship between catch and effort, $C/E = $ constant.) In order to control fishing properly, the tax would

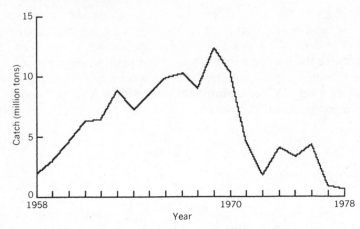

FIGURE 2.8 Total catches of Peruvian anchoveta, 1959–1978.

therefore have to be adjusted according to the current biomass level X_t—which is itself difficult to assess with accuracy, particularly for populations that experience natural fluctuations of abundance (see Fig. 2.9). It seems that catch *quotas* would therefore be indispensable in the management of a type IV fishery, although taxes could be used in conjunction. Such questions are considered in greater detail in Chapter 4.

Type III concentration profiles lead to management implications that are similar to, if less severe than, type IV. Bionomic open-access equilibrium $\bar{X}$ is positive, since $c(X)$ becomes infinitely large as $X \to 0$. For the logistic growth model,

$$G(X) = rX\left(1 - \frac{X}{K}\right) \tag{2.27}$$

the equilibrium at $\bar{X}$ is stable [i.e., $G(\bar{X}) > 0$]; also the stock will recover ($dX/dt > 0$) if fishing is reduced. However, there is reason to expect that the growth function for schooling species may become "depensatory" at low biomass levels (Clark 1974). This question and related considerations of stability will be taken up in Chapter 3.

Finally, the management implications of type I profiles are the reverse of those of type III and IV. In extreme cases, exploitation may be profitable only at low rates of catch relative to the maximum sustainable yield.† Such fisheries do still exist and are commonly referred to as

---

†This can also arise when the *demand* for fish falls rapidly as output increases, that is, when demand is "inelastic"—see Chapter 3.

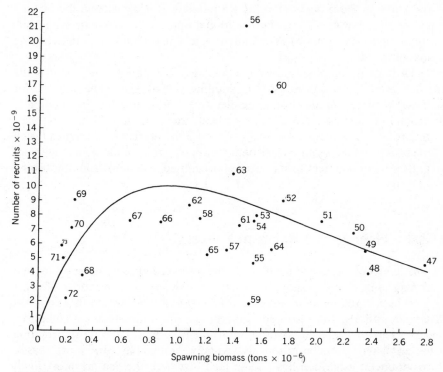

FIGURE 2.9 Recruitment of O-group North Sea herring vs. spawning stock biomass, 1947–1973; fitted by Ricker curve $S = aR\exp(-bR)$. (From Saville and Bailey 1980.)

"underutilized"—at least in a biological sense. A well-known example is Antarctic krill, where current landings, around 500,000 tonnes per annum, are small relative to the estimated MSY of 10–100 million tonnes or more per annum. In this case, exploitation is obviously limited by economic factors. However, the distribution and movement of krill are probably such that a type I profile applies. A large-scale krill fishery, if it develops, will presumably search out areas of high abundance. Krill fishing may then compete closely with the natural predators of krill, particularly the baleen whales (May et al. 1979).

What practical management measures can be suggested for the difficult—and vulnerable—case of type IV concentration profiles? Since type IV profiles usually give rise to extreme difficulty in the estimation of stock abundance, this is really a problem in resource management under uncertainty, a topic that will be discussed in Chapter 6. As will be noted there, a scientific approach to managing fisheries under uncertainty is only now in the process of being formulated. It is clear from the historical record, however, that the traditional approaches, ranging from outright laissez-faire to MSY-oriented catch regulation, are disastrously inappropriate.

Perhaps a completely different management philosophy, such as the "fail-safe" approach discussed by Holling (1978), should be adopted for these vital resources. For example, this might involve the complete prohibition of fishing in certain critical areas, or during certain times, regardless of apparent stock abundance and safe catch estimates. Such measures would require a strong, politically independent management authority, not subject to the usual shortsighted pressures from the fishing industry.

2.4   Fishing as a Random Process

The concept of concentration profile is based on an assumption that fishing units always target on the densest existing concentrations of fish.†
This in turn requires that fishermen *know* at all times where these densest concentrations are located—admittedly an unrealistic assumption. However, the opposite assumption—that fishermen have no information about stock concentration—is equally untenable. In most cases, experienced fishermen have some idea of where the fish are most *likely* to be found, and this information is used in formulating fishing strategy. Each fisherman then "updates" his prior expectations in the light of additional information obtained from "sampling" of the current fish stock. Information pertaining to fish concentrations is usually considered to be extremely valuable, a fact attested to by the many complex stratagems fishermen use to withhold it from other fishermen.

Fish concentrations are of course not uniform, even locally. Hence "sampling"—that is, locating and catching fish—is in reality a random process. (By *random* I do not mean to imply that fishing operations are completely arbitrary, but rather that fishing success is affected by partially unpredictable factors. The technical term is *stochastic*.) This has two important implications: First, the catch per unit time of individual fishing units (and hence also of "fleets" of such units) is a random variable whose *expectation* is determined by the mean density of fish, rather than being a deterministic function of density, as assumed in previous sections. As a consequence, secondly, any catch is only an

---

†This section may be postponed until Chapter 6.

indicator of local fish abundance—for example, a sequence of several good catches may indicate high abundance, or it may be due merely to a run of "good luck."

How then does the fisherman decide where and when to fish? A simple decision rule would be: (a) As long as catches are as good as can be expected, or better, stay where you are—unless you learn of better catches elsewhere; but (b) if catches are much lower than expected, try another spot. In order to give quantitative meaning to this simple rule, we shall introduce the technique of *Bayesian updating* of probabilities. (It will turn out that the simple rule is suboptimal in certain circumstances.)

Updating

Consider a single fishing "ground," with the property that there is no discernible tendency for fish to be more concentrated in one spot than another. Let X_i denote the abundance of fish occurring on the ground in year *i*. We shall suppose that X_i is a random variable with probability distribution function *f*, that is,

$$\text{pr}(X < X_i \le X + dX) = f(X)\, dX, \qquad X \ge 0 \qquad (2.28)$$

Let $\bar{X} = E\{X_i\}$ and $\sigma_X^2 = E\{(X_i - \bar{X})^2\}$ denote the mean and variance of X_i, respectively. [The probability distribution function *f* would be derived from historical data and would itself be updated (see below) as this record accumulated.]

The particular form of the function *f* is not crucial. Recruitment data often approximately fit a lognormal distribution (Hennemuth et al. 1980), but the normal distribution may be more appropriate for local stock abundance. In fact, as Swierzbinski (1981) points out, almost any two-parameter distribution $f(X; \alpha, \beta)$, defined and positive for $X > 0$, would suffice in most cases. A particularly convenient form, for reasons that will soon become apparent, is the *gamma* distribution

$$\gamma(X; \nu, \alpha) = \frac{\alpha^\nu}{\Gamma(\nu)} X^{\nu-1} e^{-\alpha X}, \qquad X > 0 \qquad (2.29)$$

The mean and variance of this distribution are

$$\bar{X} = \frac{\nu}{\alpha} \qquad \text{and} \qquad \sigma_X^2 = \frac{\nu}{\alpha^2} \qquad (2.30)$$

and the coefficient of variation is therefore $\sigma_X / \bar{X} = 1/\sqrt{\nu}$ (see this chapter's Appendix). Thus if the mean $\hat{X}$ and variance $\hat{\sigma}^2$ are known for the historical record of X_i, then the appropriate gamma distribution parameters are immediately deduced from Eqs. (2.30):

$$\alpha = \frac{\hat{X}}{\hat{\sigma}^2} \quad \text{and} \quad \nu = \frac{\hat{X}^2}{\hat{\sigma}^2}$$

Within the given fishing ground A it will be assumed that fish are, in some sense, "uniformly" distributed at random. More specifically, we shall suppose that fish occur, and are encountered, in "clumps"—which may consist of schools or other aggregations, or merely individual fish in the case of species of solitary habit. To keep the calculations reasonably simple we will also assume that the "clumps" are all of the same size. It will be convenient to measure fish abundance X in terms of the number of "clumps" in A; we will henceforth use the term *school* rather than *clump*.

The process of searching for randomly distributed objects is appropriately modelled by means of the Poisson process, one of the most frequently used models in probability (Ludwig 1974; Pielou 1977, chap. 7; Koopman 1980). The Poisson process is defined by the condition

Probability of one "event" in time $(t, t + dt) = \lambda \, dt + o(dt^2)$ (2.31)

where $\lambda =$ constant represents the expected, or average, rate of occurrence of "events" and the expression $o(dt^2)$ represents additional terms which are negligible for small dt. In the fishery setting, an "event" means an encounter with a school of fish, by fishing gear or by the searching method in use. Thus λ represents the expected rate of encounter of fish schools.

If V denotes the total volume of seawater encompassed by the fishing ground A under consideration, we may express λ in terms of the symbols introduced in Section 2.1:

$$\lambda = \frac{\epsilon e}{V} X \quad \text{(per h)} \tag{2.32}$$

where e denotes fishing effort, that is, the volume of seawater searched or screened per hour by the given "searching unit."

More generally, suppose there are E *nominal* effort units searching *independently* for fish in A. Then since

pr(one encounter by $e = aE$ searching units in time $dt) = aE\lambda \, dt$

we conclude that the *expected* catch rate is given by

$$\bar{C} = \frac{\epsilon a E X}{V}$$

$$= qEX \quad \text{schools/hr} \tag{2.33}$$

Thus the search process with nominal effort E can be modelled by a Poisson process with parameter qEX.

For the Poisson process with one searcher, Eq. (2.31), it is well known that the probability of n encounters in a time interval of length t is given by (Feller 1957)†

$$\text{pr}(n \mid \lambda) = \frac{(\lambda t)^n}{n!} e^{-\lambda t} \tag{2.34}$$

(For k independent searching units, the same formula holds, with λ replaced by $k\lambda$.) We express this as a *conditional* probability because we wish to consider the possibility that λ and hence the current stock abundance X is unknown. When the value of λ *is* known, then Eq. (2.34) applies. But we are more interested in the *inverse* problem: Given that n encounters (i.e., schools observed and presumably caught) have occurred in time t, what can we say about λ—that is, about the current stock abundance?

First, if we have no prior knowledge about the likely stock abundance in A, the best assumption is that our sample catch is representative of the entire area. This leads to the estimate

$$\hat{X} = n \frac{V}{\epsilon e t} \tag{2.35}$$

This is known as the maximum likelihood estimator (MLE) of X. Note that Eq. (2.35) is nothing other than the catch-per-unit-effort estimate of abundance.

Would an experienced fisherman use this method to estimate local fish abundance from his recent catches? Probably not! Suppose, for example, that in an area where his experience tells him that there should be good fishing at this time of year, the fisherman experiences a day or two of poor catches. He may nevertheless continue fishing for a while longer before moving off, in the hope that the poor catches were just bad luck. Of course, if catches don't improve, the fisherman will be forced to try elsewhere.

What the fisherman is doing is employing a "prior distribution" $f(X)$, which he then *updates* (no doubt quite intuitively) on the basis of his current observations. This can be expressed mathematically via Bayes's

---

†Here we ignore the fact that a school, when encountered, is *removed* from the sea—a reasonable approximation as far as the individual searcher is concerned, in most cases. See Mangel and Clark (1983) for the details when removals are taken into consideration.

theorem (see Appendix to this chapter):

$$f(X \mid n) = \frac{\mathrm{pr}(n \mid \lambda_X) f(X)}{\displaystyle\int_0^\infty \mathrm{pr}(n \mid \lambda_Y) f(Y) \, dY} \tag{2.36}$$

where $f(X \mid n)$ denotes the *conditional* probability density function for the current value of X, given that n encounters have occurred in time t, and where λ_X is given by Eq. (2.32). [The expression $\mathrm{pr}(n \mid \lambda_X)$, given in Eq. (2.34), involves the time interval t explicitly.] We refer to $f(X \mid n)$ as the *updated* probability distribution for X, in the sense that the prior distribution $f(X)$ has been combined with current fishery information, namely $n = $ number of schools actually caught, to obtain a more recent estimate $f(X \mid n)$. A numerical illustration is given in Figure 2.10.

We now see the reason for preferring the gamma distribution as our prior distribution, for if $f(X) = \gamma(X; v, \alpha)$ is substituted in Eq. (2.36), we easily obtain (see Appendix)

$$f(X \mid n) = \gamma(X; v + n, \alpha + bt), \qquad b = \frac{\epsilon e}{V} \tag{2.37}$$

In other words, the updated distribution is again a gamma distribution, but with updated parameters

$$v' = v + n, \qquad \alpha' = \alpha + bt, \qquad \text{with } b = \frac{\epsilon e}{V} \tag{2.38}$$

(If there are k independent searching units, with a total of n encounters, we get $\alpha' = \alpha + kbt$.) Because of this simple relationship, in which the updated distribution $f(X \mid n)$ has the same analytic form as the prior distribution $f(X)$, the gamma distribution is called the *conjugate* distribution (or conjugate prior) to the Poisson distribution. Because of their mathematical tractability, conjugate distributions are often used in decision theory. Nonconjugate distributions can be used when necessary, but the computations become much more difficult.

After updating, the expected stock abundance becomes

$$\bar{X}' = \frac{v'}{\alpha'} = \frac{v + n}{\alpha + bt}$$

We have $\bar{X}' > \bar{X} = v/\alpha$ if and only if

$$\frac{n}{t} > \frac{bv}{\alpha} = b\bar{X} = \frac{\epsilon e}{V} \bar{X} = \epsilon e \bar{\rho}$$

(where $\bar{\rho} = \bar{X}/V$ is the expected density of schools), which is the prior

TABLE 2.1 Symbols Used in the Updating Model

Symbol	Meaning	Units
X	Number of "schools" in A	—
$f(X)$	Prior probability density function for X	—
$\bar{X}, \sigma_x$	Mean and variance of the prior	—
ν, α	Gamma distribution parameters	—
t	Time	days
λ	Poisson distribution parameter	day^{-1}
$f(X\,|\,n)$	Conditional probability density function	
b	Expected portion of X caught per unit time	day^{-1}

expected rate of encounter of schools. In other words, if the observed encounter rate n/t exceeds the prior expected encounter rate, then the prior estimate of X is revised upward as a result of updating, and vice versa.

Also note that the coefficient of variation $1/\sqrt{\nu'} = 1/\sqrt{\nu + n}$ decreases over time as the number of schools observed increases—the information obtained by fishing reduces the uncertainty about fish abundance.

The symbols employed in the updating model are summarized in Table 2.1.

Example. To illustrate these results we consider the following simple example. Suppose

$$\bar{X} = 10{,}000, \qquad \sigma_x = 5{,}000$$

This gives $\nu = 4$ and $\alpha = 4 \times 10^{-4}$. The coefficient of variation in the prior estimate is 50%—this corresponds to expected annual fluctuations in abundance X.

Also suppose that the expected catch rate per vessel, when $X = 10{,}000$ schools, is 10 schools per day. Thus $b = 10^{-3}$. Then, if n schools are actually caught in one day's fishing, we have

$$\bar{X}' = \frac{\nu'}{\alpha'} = \frac{\nu + n}{\alpha + bt} = \frac{4 + n}{1.4} \times 10^4$$

$$\sigma_{X'} = \frac{\sqrt{\nu'}}{\alpha'} = \frac{\bar{X}'}{\sqrt{4 + n}}$$

For example:

n	$\hat{X}$	$\bar{X}'$	σ'_X	CV'
5	5,000	6,300	2,100	33%
10	10,000	10,000	2,680	27%
20	20,000	17,100	3,490	20%

The second column in the above tabulation gives the naïve "catch-per-unit-effort" estimate of abundance $\hat{X} = n/bt = n \times 10^3$. This estimate, of

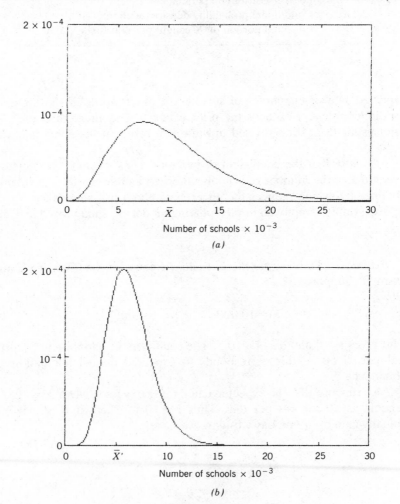

(a)

(b)

FIGURE 2.10(a, b).

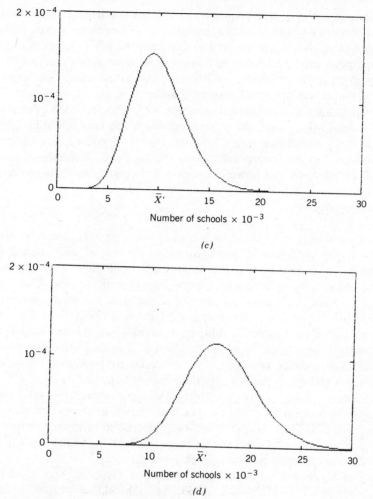

(c)

(d)

FIGURE 2.10 Gamma probability density functions: (a) prior distribution; (b)–(d) updated distributions resulting from catches of 5, 10, and 20 schools, respectively.

course, is more variable than the updated estimate $\bar{X}'$, because it ignores the experience of previous years. The original and updated probability densities, $f(X)$ and $f(X \mid n)$, are illustrated in Figure 2.10. (As a test of understanding, the reader may wish to recompute the above tablulation for the case where 10 vessels catch 50, 100, or 200 schools in a day.) Note also that the updated abundance estimate is less uncertain—the new coefficient of variation CV' decreases with n.

Management agencies frequently use rather sophisticated methods to obtain forecasts of fish stock abundance. Such forecasts, if they include variance estimates, could be used to generate the prior distribution f for the updating method. Also, in cases where the agency's estimates of observed abundance are derived from catch-effort data, the updating method could be employed to improve the accuracy of these estimates. In fact, when there exist statistical subareas with zero recorded effort (Fig. 2.1), updating (i.e., using the prior estimate) is the *only* scientific way of obtaining an abundance index for such cases. Of course, any available information can be incorporated into the estimate, including environmental correlations and historical correlation coefficients between areas.

A Decision Model

Suppose now that the fisherman has m fishing grounds to choose from. Which should he choose at the beginning of the season, and when should he decide to switch to a different ground? [The same problem faces natural predators searching for prey concentrations; see Clark and Mangel (1984). A simpler version of the problem has been extensively discussed in the statistical literature; see Gittins (1979).]

A naïve answer would be that the fisherman should first choose the ground with the highest expected catch rate. If upon fishing this ground he finds that catches are poor (i.e., his updated abundance estimate is low), he should then switch to the next best option, and so on.

Obviously, however, there are other things to consider—particularly the *costs* of transporting fishing vessels to, from, and between different fishing grounds. Thus ground B will be preferred to ground A only if the difference in expected catches is large enough to make up for the difference in the costs of fishing each ground.

Empirical studies of fishermen's behavior (Hilborn and Ledbetter 1979, Bockstael and Opaluch 1983, Swierzbinski et al. 1980) have suggested that fishermen employ additional criteria (other than expected revenues and costs), which might be categorized as "psychological costs." For example, in the British Columbia salmon fishery, Hilborn and Ledbetter (1979) found significant differences in catch rates among different fishing areas available to the same fleets. The more exposed waters generally yielded higher catch rates, suggesting that fishermen preferred to fish in sheltered areas and required significantly greater catch rates to entice them elsewhere.

In a study of the Gulf of Carpentaria (Australia) prawn fishery, Clark and Kirkwood (1979) observed that trawlers from ports in eastern Queensland (two weeks' steaming from the Gulf) seemed to require an expected "bonus" of at least $A8000 per trip *above* their opportunity cost to switch to the highly profitable banana prawn fishery in the gulf. This $8,000 presumably represents a "psychological cost" of several weeks in a remote area, far from home, not to mention the rigors and dangers of the lengthy 1500-km trip to and from the gulf.

The *relative uncertainty* associated with different fishing grounds may also be an important consideration to fishermen (Bockstael and Opaluch 1983). This uncertainty may be "universal," in the sense that fluctuations are more pronounced, and stock abundance less predictable, on ground A than on ground B, or it may be partly "individual," in the sense that a particular fisherman possesses less information about one ground than about the other. Fishermen who are averse to risks may exhibit a preference for lower, more certain catch rates over higher, more uncertain, or fluctuating catch rates (Swierzbinski, 1981, chap. 5).

We next construct a simple *decision model* for the individual fisherman (or vessel owner), who for simplicity will be assumed risk-neutral. The fisherman's objective, therefore, is to maximize his expected net income over the fishing season. Let there by m fishing grounds $A_1, \ldots, A_m$, and let $f_i(\lambda_i)$, $i = 1, \ldots, m$, now denote the fisherman's prior distribution functions for the various grounds. [We here use λ_i = expected encounter rate, rather than X_i = actual abundance, since it is λ_i that is of concern to the fisherman, who will usually not know what the relationship between λ_i and X_i is—see Eq. (2.32): $\lambda_i = (\epsilon e / V_i) X_i = b_i X_i$.] Fishing on each ground is a Poisson process with parameter λ_i, equal to the expected catch rate. (We imagine a single species of fish, the same on all grounds; however, a minor modification of the model would allow for different species, or species complexes, if these were lumped into units based on landed value.)

We also suppose that the fish stocks on the various grounds fluctuate *independently* from one fishing season to the next; that is, the random variables λ_i are independent. (Again, straightforward modification of the model would allow for correlated distributions, but the analysis becomes more difficult.)

Let us suppose that the fisherman can make a total of T fishing trips during the season and is free to select any of the m grounds for any trip. Let c_i denote the total cost of one trip to ground i, including costs of transporting the vessel and of fishing for a fixed period of time Δt, independent of the ground chosen. If he chooses ground i in the first period, his catch x_{i1} will be a Poisson random variate with (uncertain) parameter $\lambda_{i1} \Delta t$. Net revenue from the trip will be

$$\pi_{i1} = p x_{i1} - c_i \qquad (2.39)$$

Expected net revenue is therefore (a double expectation)

$$E(\pi_{i1}) = E_{\lambda_{i1}}\{E_{x_{i1}}[px_{i1} - c_i \,|\, \lambda_{i1}]\}$$

$$= E_{\lambda_{i1}}\{p\bar{\lambda}_{i1}\,\Delta t - c_i\}$$

$$= p\bar{\lambda}_{i1}\,\Delta t - c_i \tag{2.40}$$

where $\bar{\lambda}_{i1}$ denotes the prior expected catch rate

$$\bar{\lambda}_{i1} = \int_0^\infty \lambda_{i1} f_i(\lambda_{i1})\, d\lambda_{i1}$$

$$= \frac{\nu_{i1}}{\alpha_{i1}}$$

in the case of a gamma prior distribution.

What is the fisherman's optimal strategy: Which of the m fishing grounds should be selected on each trip? This question turns out to be rather subtle.

Passive Adaptive Fishing Strategies

If only one trip is possible, the optimal strategy is obvious: Select the fishing ground A_i with the highest expected net revenue $E\{\pi_{i1}\}$, as given by Eq. (2.40). This suggests the following strategy in the general case: For the first trip choose A_i to maximize $E\{\pi_{i1}\}$ as above. Then use the catch data obtained from this trip to update the prior distribution for the ground A_i that was fished. If (as a result of poor catches) this ground is no longer the most profitable (in terms of updated expectation), then select the most profitable from the remaining grounds for trip 2. In other words, the fisherman on each trip selects the most profitable appearing fishing ground, on the basis of current updated information. Such a strategy is sometimes referred to as a "passive adaptive" strategy (Walters and Hilborn 1976).

(We are tacitly assuming here that the individual fisherman obtains no new information about the fishing grounds that he does not fish. However, since our present model ignores depletion, fishermen from the other grounds would have no reason to withhold such information. In any event, the above strategy can still be used: Each fisherman updates all the f_i values for which he has new data and then selects the most profitable-appearing ground. The question of information becomes much more critical, of course, when realistic depletion effects are taken into consideration—see Mangel and Clark 1983.)

A fishing fleet in which individual vessels employ a passive adaptive fishing strategy will thus continue to search for the apparently most productive fishing ground, taking cost differentials into account. But whenever observed catches are as good as or better than expected, the fleet may fail to discover the most productive ground in some years. This leads one to suspect that improved strategies may exist.

Active Adaptive Fishing Strategies

Suppose, for example, that there are just two fishing grounds, A_1 and A_2. Suppose that $\lambda_1 = \nu_1/\alpha_1$ is *slightly* larger than λ_2, but that stock fluctuations are much less pronounced on A_1 than on A_2—that is, $\nu_1 \gg \nu_2$. Assume $c_1 = c_2$. Then the passive adaptive strategy is first to fish A_1, switching to A_2 if catches are poor.

Since $\bar{\lambda}_1 \approx \bar{\lambda}_2$, there is about a 50% chance that A_2 will not be fished at all in any given season. But since stocks on A_2 are highly variable, there is therefore about a 25% chance of missing an outstanding year on A_2 by using the passive strategy. If instead, A_2 is sampled first, the fisherman can still return to A_1 if catches on A_2 are not high; the expected loss $p(\lambda_1 - \lambda_2)$ is small from the "test" fishing. (Full details, and a numerical example, are worked out in the Appendix to this chapter.)

The second fishing strategy is called an "active adaptive" strategy. Such a strategy considers not only expected catches (and revenues), but also the value of the *information* that will be obtained from fishing. In cases of high uncertainty (small ν), this information may be sufficiently valuable to warrant some loss in expected revenue at the beginning of the fishing season.

If several vessels are fishing cooperatively, our argument indicates the importance of early-season test-fishing by some of the vessels, especially on highly variable fishing grounds (Mangel and Clark 1983). On the other hand, if vessels are competing for scarce fish, the search strategies used might be suboptimal, because of the various externalities involved. For example, the possibility of depletion of the stock (or imminent fishery closure) could lead to excessive search and fishing effort at the beginning of the season (Mangel and Clark 1983).

Conversely, if the results of searching cannot be kept secret, so that good areas are quickly invaded by competing vessels, then searching may occur at a less than optimal rate. An example is in the oyster dredge fishery of the Fouveaux Straits in New Zealand (Cranfield 1979). It is known by the fishermen that small, dense oyster concentrations exist around the edge of the vessel dredging area, but these beds are seldom sought out, because if found they must be shared with other vessels, which easily observe one another's operations. (The unfished beds are thought by biologists to be an important "refuge" of breeding stock for this closely regulated, highly profitable fishery.)

It should now be clear that searching for the most profitable fishing area is an important activity for fishermen. While this does not necessarily imply that all fishing effort will be directed toward the highest existing concentrations of fish (cost effects and uncertainty may prevent this), it does suggest that this assumption should not be an unreasonable approximation for most fisheries. Severe exceptions could arise in cases where some fishing grounds are remote from ports (or provide difficult fishing conditions) and in new fisheries in which abundance patterns are little understood. With such exceptions in mind, the concept of the concentration profile and its implications can be considered to be of practical importance in fishery modelling and management.

Appendix

Bayesian Updating

Bayes's formula is usually quoted for discrete probabilities, in the form (Feller 1957, p. 114)

$$pr(B_i \mid A) = \frac{pr(A \mid B_i)\, pr(B_i)}{\sum_j pr(A \mid B_j)\, pr(B_j)} \tag{2.41}$$

The corresponding formula for conditional probability density, Eq. (2.36), that is,

$$f(X \mid n) = \frac{pr(n \mid X)f(X)}{\int pr(n \mid Y)f(Y)\, dY} \tag{2.42}$$

is proved by the same argument. We use the elementary equations

$$pr(A \text{ and } B) = pr(A \mid B)\, pr(B) = pr(B \mid A)\, pr(A)$$

When applied to the events

$$A: \quad X \in (X_0, X_0 + dX_0), \qquad B: \quad n \text{ encounters}$$

this becomes (on letting $dX_0 \to 0$)

$$f(X_0 \mid n) = \frac{pr(n \mid X_0)f(X_0)}{pr(n)} \tag{2.43}$$

But it is obvious that

$$\text{pr}(n) = \int \text{pr}(n \mid Y) f(Y) \, dY \tag{2.44}$$

and the proof of (2.42) is complete.

The Gamma Distribution

The gamma distribution is defined by

$$\gamma(X; \nu, \alpha) = \frac{\alpha^{\nu}}{\Gamma(\nu)} X^{\nu-1} e^{-\alpha X}, \qquad X \geq 0 \tag{2.45}$$

where α, ν are positive parameters. The *gamma function* $\Gamma(\nu)$ appearing in the denominator is defined by

$$\Gamma(\nu) = \int_0^{\infty} t^{\nu-1} e^{-t} \, dt \tag{2.46}$$

It is well known, and easy to show, that

$$\Gamma(\nu + 1) = \nu \Gamma(\nu) \tag{2.47}$$

[which implies that $\Gamma(n+1) = n!$ for integer $n \geq 0$]. It follows immediately from (2.46) that (2.45) defines a bona fide probability distribution, that is,

$$\int_0^{\infty} \gamma(X; \nu, \alpha) \, dX = 1 \tag{2.48}$$

We obtain

$$\begin{aligned}
\bar{X} &= \int_0^{\infty} X \gamma(X; \nu, \alpha) \, dX \\
&= \frac{\alpha^{\nu}}{\Gamma(\nu)} \int_0^{\infty} X^{\nu} e^{-\alpha X} \, dX \\
&= \frac{\alpha^{\nu}}{\Gamma(\nu)} \int^{\infty} \gamma(X; \nu+1, \alpha) \, dX \cdot \frac{\Gamma(\nu+1)}{\alpha^{\nu+1}} \\
&= \frac{\nu}{\alpha}
\end{aligned} \tag{2.49}$$

from (2.47) and (2.48). The verification that $\sigma_x^2 = \nu/\alpha^2$ is similar.

Conjugate Updating

The derivation of Eq. (2.37) for the updated gamma prior distribution goes as follows. First, for the Poisson process we have

$$\text{pr}(n \mid X) = \frac{(\lambda_X t)^n}{n!} e^{-\lambda_X t}$$

$$= \frac{(bXt)^n}{n!} e^{-bXt} \qquad \left(b = \frac{\epsilon e}{V}\right)$$

from Eqs. (2.32), (2.34), and (2.37). Hence we obtain

$$\int_0^\infty \text{pr}(n \mid Y) \gamma(Y; \nu, \alpha) \, dY = \frac{(bt)^n}{n!} \frac{\alpha^\nu}{\Gamma(\nu)} \int_0^\infty Y^{n+\nu-1} e^{-(\alpha+bt)Y} \, dY$$

$$= \frac{(bt)^n}{n!} \frac{\alpha^\nu}{\Gamma(\nu)} \frac{\Gamma(n+\nu)}{(\alpha + bt)^{n+\nu}} \qquad (2.50)$$

by the same simple trick used in (2.49). Substituting these expressions into (2.42) and simplifying, we obtain finally

$$f(X \mid n) = \gamma(X; \nu + n, \alpha + bt) \qquad (2.51)$$

which is the desired result.

Optimal Search Strategy

We now consider the problem of optimal selection of alternative fishing grounds, as discussed in Section 2.4. For simplicity we consider the case of two fishing grounds A_1 and A_2. Let $\nu = (\nu_1, \nu_2)$ and $\alpha = (\alpha_1, \alpha_2)$ denote the parameter values for the prior gamma distributions for the two grounds.

Define $J_n(\nu, \alpha)$ as the maximum expected return (i.e., using an optimal selection policy) when there are n trips left to be completed in the given season $(n = 1, 2, \ldots, T)$. Here ν and α represent the current values of the gamma parameters, which have already been updated if one or more trips have been made this season. The method of dynamic programming (Bellman 1957) proceeds to determine the optimal policy by iteration on n. This is rather difficult, even with the aid of a computer, and we will give only a partial solution here.

For $n = 1$ we have, by Eq. (2.40),

$$J_1(\nu, \alpha) = \max_{i=1,2} E\{px_i - c_i\}$$

$$= \max_{i=1,2} \left(p \frac{\nu_i}{\alpha_i} - c_i\right) \qquad (2.52)$$

Thus for the terminal period ($n = 1$), the fisherman simply selects the most profitable ground, in terms of expected catch, using his current parameter estimates.

For $n = 2$ we can write

$$J_2(\boldsymbol{\nu}, \boldsymbol{\alpha}) = \max_{i=1,2} \left[p \frac{\nu_i}{\alpha_i} - c_i + E[J_1(\boldsymbol{\nu}', \boldsymbol{\alpha}')] \right] \qquad (2.53)$$

where $\boldsymbol{\nu}'$ and $\boldsymbol{\alpha}'$ represent the updated parameters after the first (i.e., second last) period of fishing

$$\left. \begin{array}{l} \nu_j' = \nu_j + n_j \\ \alpha_j' = \alpha_j + b \, \Delta t \end{array} \right\} \qquad (2.54)$$

where n_j is the number of schools caught on ground j (so $n_j = 0$ if $i \ne j$ since A_i is the ground actually fished in the first period). The validity of Eq. (2.53) becomes obvious on reflection, and is a special case of *Bellman's equation*

$$J_{n+1}(\boldsymbol{\nu}, \boldsymbol{\alpha}) = \max_{i=1,2} \left[p \frac{\nu_i}{\alpha_i} - c_i + E\{J_n(\boldsymbol{\nu}', \boldsymbol{\alpha}')\} \right] \qquad (2.55)$$

The expectation $E\{J_1(\boldsymbol{\nu}, \boldsymbol{\alpha}')\}$ in Eq. (2.53) is the double expectation with respect to n_i and λ_i (this is referred to as the "preposterior expectation" in the decision analysis literature). For $i = 1$, for example, it can be written explicitly as

$$E\{J_1(\boldsymbol{\nu}', \boldsymbol{\alpha}')\} = \sum_{n_1=0}^{\infty} J_1(\nu_1 + n_1, \nu_2, \alpha_1 + bt, \alpha_2) \, \mathrm{pr}(n_1)$$

$$= \left(\frac{\alpha_1}{\alpha_1 + bt} \right)^{\nu_1} \sum_{n_1=0}^{\infty} \max \left(p \frac{\nu_1 + n_1}{\alpha_1 + bt} - c_1, \, p \frac{\nu_2}{\alpha_2} - c_2 \right)$$

$$\times \left(\frac{bt}{\alpha_1 + bt} \right)^{n_1} \frac{\Gamma(n_1 + \nu_1)}{n_1! \, \Gamma(\nu_1)} \qquad (2.56)$$

from (2.50) and (2.52). This series must be evaluated numerically in most cases.

After computing the series (2.56) and the analogous series for $i = 2$, the optimal ground A_i is determined by maximizing, as in (2.53). Thus the optimal search strategy for the case $n = 2$ is found.

In *theory*, one can repeat the above procedure for $n = 3, 4, \ldots$ trips. The actual calculations, however, become extremely formidable, for reasons that become apparent as soon as one tries to organize the calculation. Fortunately, it appears that in many cases a near-optimal first-trip strategy is obtained already at the second stage, $n = 2$. Because of this, most actual implementations of the method terminate the calculation at $n = 2$. This procedure, which we refer to as the "semiactive adaptive" method (see Chapter 6), will be employed in the following example.

Example

Consider two fishing grounds A_1, A_2 with prior distributions $\gamma(\lambda_i; \nu_i, \alpha_i)$, where now $\lambda_i =$ expected (Poisson) encounter rate, in schools per day. The assumed values of ν_i, α_i, and resulting λ_i and coefficients of variation CV_i are given in the following tablulation:

	ν_i	α_i	λ_i	CV_i
A_1	4.0	0.4	10	0.5
A_2	0.08	0.01	8	3.54

A fishing trip is assumed to consist of $\Delta t = 5$ days' fishing. Thus the prior expected catch per trip is 50 schools on A_1 vs. 40 schools on A_2. The standard deviations from one year to another are, respectively, 25 schools and 142 schools. This is the situation described in Section 2.4, in which A_1 has a slightly higher average catch rate but the stock on A_2 fluctuates wildly.

For simplicity assume $c_1 = c_2$; then costs can be ignored and the price can be normalized to $p = 1$.

The passive adaptive strategy is to fish A_1 on the first trip, updating ν_1 and α_1 according to the catch made and switching to A_2 if indicated, namely if

$$\frac{\nu_1 + n_1}{\alpha_1 + \Delta t} < \frac{\nu_2}{\alpha_2}$$

(Note that, since we are using λ rather than X as our basic stock density variable, we have $b = 1$.)

Substituting our example parameters, we see that the fisherman switches to A_2 for his second trip if he catches fewer than 40 schools on A_1 on the first trip. The two-trip expected revenue (= total catch) from this strategy turns out to be 104.6 schools, which is only slightly better than the 100 schools expected for the simple strategy of always fishing A_1. [The value 104.6 is equal to the right side of (2.53) for $i = 1$, which is easily seen to correspond to the passive strategy.]

The active adaptive strategy for this example is in fact to fish A_2 in the first period; the explanation of this was given in section 2.4. This strategy gives an expected two-period catch of 119.8 schools, which is 14.5% better than the passive strategy.

Suppose, however, that the fisherman has a total of 10 trips to make. Using the passive adaptive strategy, the fisherman first fishes A_1 and switches to A_2 only if his updated estimate λ_1' falls below λ_2. Under the semiactive adaptive strategy the fisherman first fishes A_2. The following tablulation shows the corresponding 10-period expected catches for the two strategies, assuming that no further switches are made after the second trip.

	Passive (A_1)	Semiactive (A_2)
2-period	104.6	119.8
10-period	541.5	758.0

Note that the passive strategy (A_1) increases the expected catch, relative to the nonadaptive strategy, by 8%, whereas the semiactive strategy (A_2) yields over a 50% increase. This difference represents the additional (expected) *value of information* obtained by using A_2.

Neither of the strategies considered allows for the possibility of further updating and switching after period 2. Such a switch would in fact be indicated only if the information obtained from trip 2 "contradicted" the updated estimates calculated after trip 1. This is certainly possible, and it leads to somewhat larger figures than those shown in the tabulation. However, the full calculation is not attempted here.

3 Deterministic Single-Species Models

Modelling of a complex field necessitates the development of a large variety of different models. Each particular model will focus on one or more aspects of the real-world system, abstracting from most of its complexity. The exact mathematical form of any given simple model can be modified in various ways in order to test the robustness of its predictions. Also, simple models can be combined and cross-effects investigated. These procedures of variation and synthesis are vital to the development of broad theories appropriate for understanding a complex field and for formulating management policy.

In this chapter we discuss several modifications of our basic dynamic fishery model. It is significant that this one simple model can serve as the basis for investigating a wide variety of important economic phenomena. From the biological viewpoint, however, our basic "general production" population model is quite oversimplistic, ignoring as it does age structure, multispecies interactions, and environmental factors. These features will be introduced in later sections, but first we concentrate on economic complications.

3.1 Parameter Shifts and Myopic Decision Rules

The dynamic Gordon–Schaefer model (rewritten below) contains a number of biological and economic parameters, each of which was assumed in Chapter 1 to remain constant. We now ask: What happens if these parameters shift over time?

For the case of open access, parameter shifts cause little theoretical difficulty, unless they occur more rapidly than the corresponding adjustments in fishing effort. When *fixed* costs are taken into consideration, however, such adjustments in effort—particularly large-scale adjustments—may indeed become sluggish. This important topic will be taken up in Section 3.3, but for the present we shall continue to ignore fixed costs.

In the case, then, that effort adjustment is instantaneous, the open-access fishery will simply track the bionomic equilibrium $\bar{X} = \bar{X}_t$ given by the zero-profit condition

$$c_t(\bar{X}_t) = p_t \tag{3.1}$$

where both price p_t and unit harvest cost $c_t(X)$ may vary over time. The *sensitivity* of $\bar{X}$ to price and cost shifts will depend on the form of the cost function $c(X)$. As noted in Chapter 2, for example, a type IV concentration profile results in a *discontinuous* relationship between $\bar{X}$ and the price/cost ratio p/c: The bionomic equilibrium jumps suddenly from $X = K$ to $X = 0$ when p/c exceeds a threshold value [see Eq. (2.25)]. (An instantaneous jump in biomass X of course implies an *infinite* effort adjustment in the continuous-time model. Fixed-cost considerations, however, change this prediction to a finite but possibly rapid rate of depletion once p/c surpasses the threshold level.)

One important prediction of the theory is that, in an unregulated open-access fishery, price increases, or technological improvements that result in increased efficiency and decreased effort costs, will in the *long* run have a *deleterious* effect on the fishery, leading to increased levels of fishing effort and greater levels of stock depletion. The situation is all the more insidious because the *initial* effect of such changes will appear beneficial to the fishing industry, especially (in the case of technological improvements) to those entrepreneurs who first adopt them. Indeed, such developments may even appear to constitute a solution to the long-standing problem of poor economic performance of the fishery. Without appropriate regulation, however, such benefits can only be ephemeral.

The Sole-Owner Model

The next question to be considered is how the sole owner of a fishery would react to shifts in prices and costs or other parameters. Clearly his behavior will be influenced by the degree to which he is able to forecast future changes. A realistic analysis would treat such changes as uncertain, but here we will restrict our consideration to two extreme cases: completely unexpected changes, on the one hand, and perfectly predicted changes on the other.

Unexpected parameter shifts are easily analyzed. The sole owner operates on the basis of current parameter values alone, computing the optimal stock level X^* from the marginal production rule, Eq. (1.26) and adjusting his harvest policy accordingly when parameters shift. The direction of change in X^* resulting from a positive shift in one of the parameters (p, c, δ, q, r, K) of the Gordon–Schaefer model is easy to determine from Eq. (1.26); the results are shown in Table 3.1. For example, an unexpected increase in price p leads the sole owner to harvest the resource more intensively than before, thus reducing X^*. The quantitative sensitivity of X^* to the parameter values can readily be calculated from Eq. (1.26) for any particular set of numerical parameter values. The relative importance of the different parameters can also be clarified via nondimensionalization of Eq. (1.26)—see Clark (1976a, pp. 45–47). (Table 3.1 is also valid for the generalized production model, with arbitrary concentration profile.)

Next consider the case of perfect anticipation; the parameters of the model therefore become explicit, *known* functions of time t. The mathematical optimization problem now becomes [see Eqs. (1.19)–(1.21)]

$$\underset{\{E_t\}}{\text{Maximize}} \int_0^\infty \alpha_t [p_t q_t(X_t) X_t - c_t] E_t \, dt \qquad (3.2)$$

TABLE 3.1   Qualitative Dependence of Optimal Biomass X^* on the Model Parameters

	p	c	δ	q	r	K
Change in X^*	$-$	$+$	$-$	$-$	$+$	$+$

subject to

$$\frac{dX_t}{dt} = G(X_t, t) - q_t(X_t)E_tX_t \tag{3.3}$$

where E_t satisfies

$$0 \le E_t \le E^{\max} \tag{3.4}$$

The factor α_t in Eq. (3.2) is the *discount factor* corresponding to a varying discount rate δ_t; it is given by

$$\alpha_t = \exp\left(-\int_0^t \delta_s \, ds\right) \tag{3.5}$$

(*Proof*: If V_t = accumulated value of an initial deposit V_0 at variable compound interest rate δ_t, then $dV_t/dt = \delta_t V_t$. Hence $V_t = V_0 \exp (\int_0^t \delta_s \, ds)$, and Eq. (3.5) follows from this. Note that Eq. (3.5) reduces to the usual expression $\alpha_t = e^{-\delta t}$ if $\delta_t = \delta$ = constant.)

Note the inclusion of a constraint on maximum effort in Eq. (3.4); this is obviously realistic, although we shall for the time being treat $E^{\max}$ as exogenous. The problem of determining $E^{\max}$ itself in an optimal manner is discussed in Section 3.3.

To simplify the notation we now write net revenue flow in (Eq. 3.2) as

$$\pi_t = [p_t q_t(X_t) X_t - c_t]E_t$$
$$= [p_t - c_t(X_t)]H_t$$
$$= \phi_t(X_t)H_t \tag{3.6}$$

where

$$\phi_t(X_t) = p_t - c_t(X_t) \tag{3.7}$$

$$c_t(X_t) = \frac{c_t}{q_t(X_t)X_t} \tag{3.8}$$

and where H_t denotes the harvest (catch) rate.

The optimal fishing policy can now be described (mathematical details of the derivation are given in the Appendix to this chapter). There exists a *time-varying optimal biomass level* $X_t = X_t^*$ determined by the equation

$$\frac{\partial G}{\partial X_t} + \frac{G \, \partial \phi_t/\partial X_t}{\phi_t} = \delta_t - \frac{\partial \phi_t/\partial t}{\phi_t} \tag{3.9}$$

Equation (3.9) is a time-dependent version of the golden-rule equation for the unknown X_t; the solution $X_t = X_t^*$ is in general therefore a function of time t. (In the terminology of control theory, X_t^* is called the *singular solution* of the optimization problem.) Without going into details, we shall henceforth suppose that Eq. (3.9) is *uniquely* solvable for $X_t = X_t^* \geq 0$. This is easily seen to be the case for the time-dependent Gordon–Schaefer model, in which

$$G(X_t, t) = r_t X_t \left(1 - \frac{X_t}{K_t} \right)$$

and $q_t(X_t) = q_t$ is independent of X_t. (Cases of multiple solutions can arise, but the analysis then becomes quite complicated, especially for the time-dependent case; see Spence and Starrett 1975.)

The reader will have observed the similarity between Eq. (3.9) and the "fundamental equation," Eq. (1.26), for the optimal biomass X^* in the time-independent model of Chapter 1. In fact, the two equations are identical except for the presence of one additional term, $(\partial\phi_t/\partial t)/\phi_t$, in the new equation (and of course the appearance of partial rather than ordinary derivatives). The bioeconomic significance of each term in Eq. (1.26) was explained in Chapter 1—for example, $\partial G_t/\partial X_t$ represents the marginal productivity of the resource stock X_t. The new term

$$\frac{\partial\phi_t/\partial t}{\phi_t} \tag{3.10}$$

represents the relative rate of increase of harvest profitability ϕ_t over time, which must then be subtracted from the current discount rate δ_t. This "correction" is especially obvious in the case that price p_t is the only time-dependent parameter. Equation (3.9) then becomes

$$G'(X_t) - \frac{c'(X_t)G(X_t)}{p_t - c(X_t)} = \delta - \frac{\dot{p}_t}{p_t - c(X_t)} \tag{3.11}$$

that is, the discount rate δ is "corrected" by the relative rate of (net) price increase. [Note: $\dot{p}_t$ denotes the time derivative of p_t.]

This result makes good sense economically, but the situation is more complicated than it may appear. Suppose, for example, that at time t_0 the price level, which was formerly constant, begins to rise at a constant exponential rate $\dot{p}_t/p_t = \beta$. This has *two* effects on the optimal biomass level X_t^*. First, at time t_0 the right side of Eq. (3.11) jumps downward. Consequently X_t^* jumps to a higher level—see Figure 3.1(*a*). Second, for $t > t_0$ the "marginal stock effect" term on the left side of Eq. (3.11) decreases, since p_t is increasing. This causes X_t^* to decrease over time for $t > t_0$. The entire singular path X_t^* is shown in the figure.

But now observe that this singular path is not "feasible"—no effort input $E_t \geq 0$ can bring about a discontinuous *increase* in the biomass

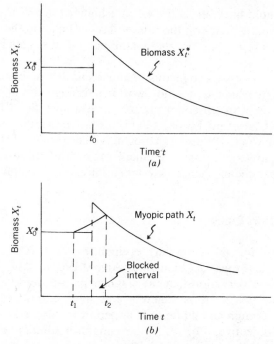

FIGURE 3.1 Optimal biomass level under price changes: (a) the myopic path X_t^*; (b) optimal feasible biomass level—fishing ceases during the "blocked interval" from t_1 to t_2.

level X_t. The best that can be done is to set $E_t = 0$ (stop fishing), allowing X_t to increase at its natural rate $\dot{X}_t = G(X_t)$. In fact, it can be shown (see the chapter Appendix) that the optimal fishing policy is indeed to stop fishing for a time interval $t_1 \leq t \leq t_2$ surrounding the discontinuity at time t_0 [Figure 3.1(b)] and then to resume harvesting along the singular path X_t^* for $t > t_2$. The appropriate level of fishing effort E_t^* is thus given by

$$E_t^* = \begin{cases} \dfrac{G(X_0^*)}{q(X_0^*)X_0^*} & \text{for } t < t_1 \\[2mm] 0 & \text{for } t_1 \leq t \leq t_2 \\[2mm] \dfrac{G(X_t^*) - \dot{X}_t^*}{q(X_t^*)X_t^*} & \text{for } t > t_2 \end{cases} \qquad (3.12)$$

This policy also makes good sense economically—and even describes observable behavior of resource stock owners (perhaps not fishery owners). Namely, the owner forsees an imminent price rise. Obviously, rather than exploiting and selling his stock at the current price level p_0, it pays the owner to hold back and sell later at a higher price. The appropriate hold-back period depends on the rate of discount and on the growth rate of the stock, as well as on the magnitude of the anticipated price increase.

The effects of other exogenous changes can be analyzed by similar techniques. The reader may enjoy working through the implications of rapid alterations in the discount rate δ_t, for example. See Clark (1976a, chap, 4) and Clark and Munro (1975, 1978) for further applications; Clark and Munro (1978) discuss cases in which resource extinction may arise and demonstrates that "optimal" extinction becomes less likely when future parameter changes are taken into consideration.

Myopic Decision Rules

Equation (3.9) for the (singular) optimal biomass X_t^* is sometimes referred to as a *myopic* decision rule (Arrow 1964).† This terminology stems from the observation that the singular path is determined entirely by the *current* values of parameters and their current time rates of change. Even though an infinite time horizon was adopted in the original optimization objective, Eq. (3.2), the (singular) solution requires only local-time data. Myopic decision rules thus provide a vast simplification in terms of data requirements.

However, as seen from our price-change examples, cases can well arise in which the myopic rule becomes infeasible because of *constraints* on the policy variable (E_t in our model). In such cases optimal management *does* usually require foreknowledge of future parameter shifts, in order to decide when to begin diverging from the myopic path. The general rule (Arrow 1968) is that a digression from the myopic path must begin *before* that path becomes infeasible (Fig. 3.2). There ensues a "blocked interval" (blocked by the constraint), during which the myopic path is not followed and the control variable (here E_t) assumes a constrained value (here either 0 or E^{max}). Let us note also the *premature switching* characteristic of blocked intervals—the optical switch to a blocked interval always occurs *before* the myopic path actually becomes infeasible.

As a final example, consider the following (which will prove useful later—see Section 3.3). Assume, in our fishery model, that all parameters remain constant over time, but suppose that

---

† The word *myopic* is used here in a different sense than in Section 2.4.

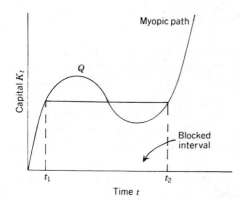

FIGURE 3.2 Myopic path and blocked interval for an investment problem. The constraint condition is $K_t \geq 0$. The first switch, which occurs at t_1, is "premature" since the myopic path could be followed up to point Q.

$$E^{\max} < \frac{G(X^*)}{q(X^*)X^*} \qquad (3.13)$$

that is, effort "capacity" $E^{\max}$ is too small to permit sustained harvesting at the constant optimal (myopic) biomass level X^* as given by Eq. (3.9). What is the optimal effort policy?

If the initial biomass X_0 is greater than X^* the optimal harvest policy is clearly to set $E_t = E^{\max}$ ad infinitum: Harvest at the maximum possible rate. The fish biomass level will converge to an equilibrium above X^*, and no blocked interval arises.

On the other hand, if $X_0 < X^*$, the optimal policy is to refrain from fishing ($E_t = 0$) until some time t_1 and then to switch to maximum effort $E_t = E^{\max}$ for $t > t_1$. By the premature switching rule, the switch occurs *before* X_t has grown to the myopic level X^*. Nevertheless, X_t eventually reaches a long-run equilibrium at a level above X^*, as before.

The determination of the optimal switching times $t_1, t_2, \ldots$ in blocked-interval problems is straightforward; see Arrow (1968) or Clark (1976a).

Myopic decision rules, while easy to treat analytically, obviously involve extreme assumptions. Wherever the fish stock X is below the optimal level X^*, for example, the myopic rule requires zero harvesting. Even if temporary, zero harvesting may have undesirable consequences for fishermen and processors whose livelihood depends on a regular supply of fish. Consumers may also be adversely affected unless substitute foods are available at competing prices.

On the other hand, harvesting at a high rate when $X > X^*$ may flood both processing facilities and markets. Dealing with these questions requires the use of nonlinear models, which we study in the next section.

3.2 Demand Schedules: Nonlinearity

The theory of myopic decision rules discussed in the previous section is especially simple and appealing. Unfortunately, however, it is restricted to a special class of dynamic optimization problems, namely those in which the control variable (here, E_t) appears *linearly* throughout the model.† It is easy to see that this is the case for the model analyzed in the previous section. But this linearity assumption has been adopted at some cost in terms of unrealistic assumptions as well as in terms of the form of the resulting optimal policy.

In particular, in order to obtain a linear model, we have been forced to assume that the price p of landed fish is independent of the catch rate H. This assumption is usually considered to be reasonable for small fisheries delivering their catches to a large market—the fishermen (or sole owners) are then "price takers," unable to affect market price by raising or lowering the supply of fish.

Many fisheries, however, face small local markets where price may be highly sensitive to the current supply of fish. This sensitivity will be especially noticeable for "luxury" species, which are not readily substitutable. On the other hand, some fisheries are so large that their output directly affects prices. The collapse of the Peruvian anchoveta fishery in 1973, for example, affected prices of fish meal and its substitutes.

Demand Schedules

The *demand schedule* for a given product relates market price p to the supply rate Q. We will write $p = P(Q)$ for the "inverse demand function"; it is assumed that this function is decreasing: $dP/dQ < 0$. The expression

$$e = -\frac{1}{Q}\frac{dQ}{dP} = -\frac{1}{Q}\left(\frac{dP}{dQ}\right)^{-1} \tag{3.14}$$

is called the *elasticity* of demand. It represents the percentage change in supply Q corresponding to a 1% change in price p. Thus our former assumption $p = $ constant (i.e., $dP/dQ = 0$) is equivalent to an assumption of *infinite elasticity* of demand.

Consider first the case of a *monopolist*, who controls the entire output

---

†This applies to differential equation models; a slightly different but analogous requirement applies to the case of difference equations (discrete-time models)—see Section 3.6 and the Appendix to this chapter.

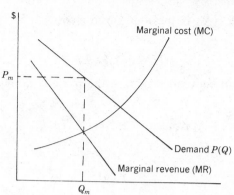

FIGURE 3.3 Supply-and-demand model: monopoly solution. The monopolist equates marginal revenue MR to marginal cost MC, thus producing output Q_m and selling at price p_m.

of some good Q. We have (note that these are *rates*)

$$
\begin{aligned}
\text{Total revenue:} \quad & \text{TR} = P(Q)Q \\
\text{Total cost:} \quad & \text{TC} = C(Q) \\
\text{Monopoly "profit":} \quad & \pi = P(Q)Q - C(Q)
\end{aligned}
$$

where $C(Q)$ denotes the monopolist's total cost of production. The profit-maximizing monopolist selects a supply rate Q that maximizes profit π, thus implying that marginal revenue equals marginal cost:

$$
\frac{d}{dQ}[P(Q)Q] = \frac{d}{dQ}C(Q) \tag{3.15}
$$

The situation is shown in Figure 3.3.

The simple textbook model involves no constraint on supply Q; it therefore requires modification for the case of a natural resource monopoly, where supply is limited by nature. The dynamic Gordon–Schaefer model is easily modified to describe the monopolist sole owner of a fishery resource (Clark and Munro 1975).

For simplicity, assume that the supply of fish equals the rate of catch. (This ignores processing and storage of fish—see Section 3.4.) The monopolist's profit flow is then given by

$$
\pi_t = \pi(H_t, X_t) = P(H_t)H_t - c(X_t)H_t \tag{3.16}
$$

where, as before, $c(X)$ is the unit harvest cost.† The catch rate H_t is given as usual by

$$H_t = q(X_t)X_tE_t \qquad (3.17)$$

The monopolist attempts to maximize the present value of his resource profits:

$$\text{Maximize} \int_0^\infty e^{-\delta t}\pi(H_t, X_t)\, dt \qquad (3.18)$$

subject to

$$\frac{dX_t}{dt} = G(X_t) - H_t \qquad (3.19)$$

$$0 \le E_t \le E^{\text{max}} \qquad (3.20)$$

The above optimization problem can be solved "analytically" only provided the underlying parameters are assumed to remain constant *over time*; the notation employed in Eqs. (3.16)–(3.20) reflects such an assumption. As in Chapter 1, there again exists an optimal equilibrium biomass level X_m^* (for the monopolist), which is now determined by the equations

$$G'(X) + \frac{\partial\pi/\partial X}{\partial\pi/\partial H} = \delta \qquad (3.21)$$

$$H = G(X) \qquad (3.22)$$

Upon writing out the indicated partial derivatives, using Eq. (3.16), we obtain

$$G'(X) - \frac{c'(X)G(X)}{\text{MR} - c(X)} = \delta \qquad (3.23)$$

where MR denotes marginal revenue:

$$\text{MR} = \frac{d}{dH}[P(H)H] = P(H) + H\frac{dP}{dH} \qquad (3.24)$$

[The derivation of Eq. (3.21) is outlined in the Appendix to this chapter.]

Note the strict similarity between Eq. (3.23) and our fundamental law, Eq. (1.26)—indeed, (3.23) immediately reduces to (1.26) for the case of infinitely elastic demand $(dP/dH = 0)$. As in the simple textbook model, the monopolist's optimal (long-run) supply level is based on marginal revenue MR, rather than price p.

---

†Following the rule of introducing complications one at a time, we here continue to assume that fishing *costs* are linear in H. See Section 3.7.

What about short-run supply? How does the monopolist optimally harvest a fishery resource when the current biomass level X_t differs from the long-run optimum X_m^*? We shall return to this question after a brief digression to consider the question of social welfare.

Socially Optimal Fishery Supply

It is a basic tenet of welfare economics that the monopolist performs in a socially undesirable manner. Economic texts (e.g., Samuelson 1980) argue that social welfare is maximized when the level of supply Q is such that *price* equals marginal cost, and not when marginal revenue equals marginal cost, as in the case for the monopolist. A purely competitive market does equate price to marginal cost and hence achieves the socially optimal level of production. The result is indicated graphically in Figure 3.4. The monopolist charges an excessive price $p_m > p_s$ for his product and hence puts less of the product on the market than the social optimum: $Q_m < Q_s$.

Putting this in mathematical terms, the *social utility* of consumption Q is defined by

$$U(Q) = \int_0^Q P(q)\, dq \tag{3.25}$$

Thus *marginal* utility $U'(Q)$ equals market price $P(Q)$. Net social welfare $U(Q) - c(Q)$ is therefore maximized when price equals marginal cost.

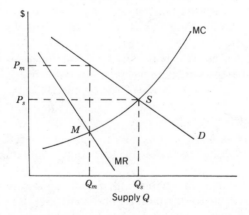

FIGURE 3.4 The social optimum (S) vs. the monopoly solution (M), in traditional supply-and-demand theory.

In the fishery setting, the net flow of social welfare corresponding to a fish supply rate H_t is given by

$$\pi_s(H_t, X_t) = U(H_t) - c(X_t)H_t \qquad (3.26)$$

Here $c(X_t)$ denotes unit social harvesting cost, which for simplicity we assume equal to private harvesting cost (see Clark 1976a, Section 3.4). The discounted present value of net social welfare is therefore given by

$$\int_0^\infty e^{-\delta_s t}[U(H_t) - c(X_t)H_t]\, dt \qquad (3.27)$$

where δ_s designates the "social rate of discount." By the same calculation used in obtaining (3.23), we then derive the equation

$$G'(X) - \frac{c'(X)G(X)}{P(H) - c(X)} = \delta_s \qquad (3.28)$$

for the *socially* optimal equilibrium biomass level $X = X_s^*$. Note that this is also the optimal biomass level for a sole owner facing a purely competitive market at the constant price $p = P(H)$.

Equation (3.28) differs from (3.23) in two respects: marginal revenue MR is replaced with price $P(H)$, and the monopolist's discount rate δ_m is replaced with the social discount rate δ_s. Ignoring, first, the possible difference between δ_m and δ_s, we see that [since MR $< P(H)$] the socially optimal biomass level is *less* than the monopolist's optimum:

$$X_s^* < X_m^* \qquad \text{if} \qquad \delta_s = \delta_m \qquad (3.29)$$

This reflects the fact that the monopolist "underproduces" relative to the social optimum. The resource monopolist is *more* conservative than is socially optimal (cf. Hotelling 1931)—provided he uses a socially optimal rate of discount.

It has often been argued, however, that discount rates in the private sector tend to be greater than socially desirable rates—especially in natural resource situations (Ramsay 1928, Solow 1974, Page 1976). If so, this will motivate the monopolist to be less conservative, that is, to exploit the stock more intensively than socially desirable. The two effects, monopoly and discounting, act in opposite directions on X_m^*, so that the final outcome, relative to the social optimum X_s^*, is not determined.

A historical example involving both phenomena is provided by the pelagic whaling industry (Clark and Lamberson 1982). Before World War II, regulation of whaling consisted solely of a limitation on total catches, with the aim of preventing an oversupply of whale oil. Without these arrangements, the whale stocks would probably have been depleted more rapidly, and both prices and profits would have been smaller than was the case. Depletion of whale stocks was not seen as a problem at this time.

But even after the war, when the severity of depletion was becoming apparent, most whaling nations displayed little interest in curtailing overall catches in order to preserve whale stocks. Thus, although the ability to limit catches in order to preserve profits suggests that the whaling nations possessed some degree of controllability, this did not appear to imply any noticeable motivation toward conservation over the long run. But this is clearly oversimplified—the whaling "cartel" may have felt strong enough for short-term, but not long-term measures, for example. And eventually, conservation measures were adopted.

Dynamics of Adjustment

Our discussion so far has been mainly restricted to *equilibrium* solutions X^* of the sole-owner model. These are easy to understand and to analyze, because of the simplicity of their characterizing equations [see Eqs. (1.26), (3.23), and (3.28)]. There remains the question of determining the optimal rate of adjustment from an initial biomass level X_0 different from X^*. Adjustments in biomass occur, of course, through changes in harvest rate (and effort), and these are obviously of economic importance. (Natural fluctuations also occur, of course, but we are putting off this question until Chapter 6.)

Let us return for a moment to the model of Chapter 1; in that model both the catch rate H_t and the rate of net revenue π_t are *linearly* dependent upon effort E_t:

$$H_t = qX_tE_t, \qquad \pi_t = [p - c(X_t)]E_t$$

This linearity property implies that adjustments in biomass should always be made as rapidly as possible. The reason should be fairly obvious: Economic losses occur whenever biomass X differs from X^*. Any delay in adjustment means that these losses are incurred for longer periods than necessary. Because of the linearity, the costs and benefits associated with adjustment are the same for rapid as for slow adjustments. Therefore adjustments in biomass should take place as rapidly as possible. Mathematically, the rule is

$$E_t = \begin{cases} E^{\max} & \text{if} \quad X_t > X^* \\ 0 & \text{if} \quad X_t < X^* \end{cases}$$

where $E^{\max}$ denotes the maximum possible effort level. The corresponding time profiles of catch and biomass level are illustrated in Figure 3.5.

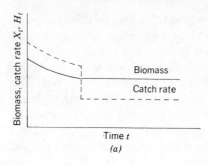

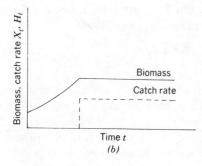

FIGURE 3.5 Optimal catch rate and biomass profiles during the adjustment phase, for the linear model: (a) initial stock level $X_0 > X^*$; (b) initial stock level $X_0 < X^*$. The optimal catch rate shifts discontinuously to H^* when X^* is reached.

The optimal effort policy for the linear model has the property that the catch rate shifts suddenly when X reaches the optimum level X^*. This may seem unrealistic—it is usually thought that a steady supply of a product is superior to such shifts in supply. (In forestry, the downward shift from harvesting virgin forests to sustained-yield harvesting, referred to as the "fall-down effect," has been the subject of intensive discussion.)

The phenomenon of sudden shifts in H_t is strictly dependent on the linearity property of the model; it disappears, for example, when finite demand elasticity $[P(H_t) \neq \text{constant}]$ is introduced into the model. This is

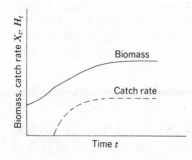

FIGURE 3.6 Optimal catch rate and biomass profiles during the adjustment phase for a nonlinear model with $p = P(H_t)$. Note that, in contrast to Figure 3.5, the catch rate increases smoothly to its long-run equilibrium value.

illustrated, for the case of an initially depleted stock $X_0 < X^*$, in Figure 3.6. In this case the catch rate must be small initially, in order to permit stock recovery. But it is not optimal to maintain an absolute moratorium until recovery is complete (although a temporary moratorium may well be appropriate, as illustrated, if the stock is severely depleted). Rather, the harvest rate should be gradually increased up to the optimum sustainable yield $H^* = G(X^*)$. The exact form of the optimal catch H_t (or effort E_t) depends on the various parameters of the model; it can be obtained only by solving a certain system of differential equations (see Clark (1976a, chap. 5). A similar continuous adjustment applies to the case in which $X_0 > X^*$.

These results have considerable practical significance. The option of completely closing down a depleted fishery, while it may lead to the most rapid possible recovery of the stock, can have extremely unpleasant consequences both for the fishing industry and for fish consumers. Reducing catches sufficiently to ensure *any* degree of recovery can be pretty traumatic, but is necessary in order to ensure improved catches in the future. A complete moratorium, however, can only be justified under special circumstances. In some cases, quite low catch levels can still provide fishermen and processors with satisfactory incomes, at least temporarily, as prices respond to the short supply. In a severely overexpanded fishery, however, it may not be possible to meet the income requirements of all participants. (Additional economic reasons for the nonoptimality of complete moratoria will be discussed in the next section.)

For a naturally *fluctuating* fish stock (see Chapter 6), the myopic model implies that optimal harvesting should attempt to maintain a constant "target" biomass level X^*, regardless of stock fluctuations. Such a policy can obviously lead to severe fluctuations in annual catches. As the present analysis indicates, however, such extreme fluctuations may be economically and socially undesirable. This is not merely an academic matter, especially in view of prevalent fishery management practice, in which a fixed "target escapement" is often considered the ideal.

Nonlinear Effort Costs

We conclude this section by discussing briefly the implications of a nonlinear cost-of-effort function $C(E)$. We shall assume that the marginal cost of effort is increasing:

$$C'(E) > 0, \qquad C''(E) \geq 0 \tag{3.30}$$

This somewhat ad hoc formulation allows us to assume nonhomogeneity of harvesting units, without explicitly introducing additional structure into the model (see Chapter 4). Net revenue flow to the fishery is then given by

$$\pi = pH_t - C(E_t)$$

$$= pqX_tE_t - C(E_t) \tag{3.31}$$

(For simplicity we revert to the case of a constant, exogenous price p.)

Bionomic equilibrium in the open-access fishery is now characterized by the condition

$$\frac{\partial \pi}{\partial E} = pqX - C'(E) = 0 \tag{3.32}$$

since effort will expand to the level at which the *marginal* unit of effort earns zero net revenues (Copes 1972; Anderson 1977; this question will be discussed more fully in Chapter 4).

Let us write Eq. (3.32), relating effort and biomass in the open-access fishery, in the form

$$C'(E) = pqX \tag{3.33}$$

It can be shown (see this chapter's Appendix) that the optimal effort level is characterized by the equation

$$C'(E) = (p - \lambda)qX \tag{3.34}$$

where λ is a *positive* parameter, usually referred to as the *shadow price* of the resource stock. The open-access fishery thus exerts excessive effort at every biomass level X—and hence depletes the resource relative to the optimum. These predictions are in agreement with the original Gordon–Schaefer model. It is no longer the case, however, that total fishery net revenues are zero at bionomic equilibrium, but rather that marginal revenues are zero. Fishing units whose effort costs are less than those of the marginal (i.e., least efficient) units will enjoy positive returns at equilibrium. This latter prediction seems to be borne out by the recognized phenomenon of the "highliner"—a fisherman who consistently performs better than average and thus regularly enjoys positive net revenues.

The comments made earlier regarding gradual vs. most rapid adjustments to equilibrium also apply in the case of nonlinear costs. However, a more direct and significant relationship between costs and adjustment policy arises from the consideration of *fixed costs*. We now turn to this question.

3.3   Investment and Fixed Costs

In all the models studied so far, fishing costs have been assumed to be directly related to fishing effort (and, except briefly in Section 3.2, to be linearly related to effort). This assumption may be quite unrealistic.

Table 3.2 lists per vessel costs of fishing for a small scallop fishery in New Zealand; these data are typical for a small, day fishery of this type, but many fisheries are much more capital-intensive than this.

It would take us too far afield to attempt to describe the cost categories listed in the table in any detail, but it is evident that the assumption "cost proportional to effort" is a serious oversimplification. Roughly 25% of the costs listed in Table 3.2 are classified as fixed costs—costs that are incurred regardless of the amount of fishing actually undertaken.

We now introduce a modification to our basic fishing model, incorporating both variable and fixed costs (Clark et al. 1979). We continue to assume that *variable costs* of fishing are directly proportional to

TABLE 3.2   Per Vessel Costs and Earnings of Sample Vessels in the Southern Scallop Fishery of New Zealand, 1976[a]

Variable costs	
Crew remuneration	$ 6,844
Ship's stores	901
Fuel and oil	2,628
Repairs and maintenance	4,451
Shore expenses	1,153
Total:	15,977
Fixed costs	
Insurance	1,315
Depreciation	1,864
Administration expense	870
Interest	1,035
Total fixed costs	5,084
Receipts from sales	24,734
Operating income	3,673
Value of assets	19,599

Source: New Zealand Fishing Industry Board (1978).
[a]Figures in New Zealand dollars.

fishing effort E_t. Except for the category "crew remuneration," this assumption seems reasonable for the variable costs listed in Table 3.2. [Crew remuneration, in many fisheries, consists mainly of *shares* of the catch rather than predetermined wages for skippers and crews. For an economic analysis of the share system in fisheries, see Sutinen 1979.]

Fixed costs are assumed to consist of interest and depreciation on invested capital. (Insurance costs are irrelevant in a deterministic model, but could also be taken as proportional to capital value in the stochastic case.)

Sole-Owner Model

We begin by studying the case of the sole owner of a fishery. Denote by $K = K_t$ the value of fixed capital assets which the fishery owner possesses at time t. For simplicity we shall imagine that this capital consists entirely of fishing vessels plus permanent fishing gear (winches, radar sets, and the like), but in practice shore installations could also be included. The dynamics of capital adjustment are modeled by the equation

$$\frac{dK_t}{dt} = I_t - \gamma K_t \tag{3.35}$$

where I_t denotes the *rate of investment* in capital and $\gamma = \text{constant} \geq 0$ denotes the *depreciation rate*. The rate of investment I_t is subject to choice by the owner, under certain constraints to be discussed below.

The amount of fishing effort that the owner is capable of exerting is taken to be proportional to the size of his fishing fleet or, in other words, to his capital K_t. It is convenient to take, as the unit of K, 1 SFU; we thus have

$$E^{\max} = K_t$$

and effort E_t is therefore constrained by

$$0 \leq E_t \leq K_t \tag{3.36}$$

Introducing capital K_t into the model thus provides us with a natural meaning for the symbol $E^{\max}$, which previously had only an ad hoc significance.

Let c_K denote the cost of capital ($/SFU). Investment at the rate I_t (SFU/year) thus incurs a cost at the rate of $c_K I_t$ ($/year). Net revenue flow to the fishery owner is given by

$$\pi_t = [pq(X_t)X_t - c]E_t - c_K I_t \tag{3.37}$$

As before, we hypothesize that the owner will attempt to maximize his net present value:

$$\underset{I_t, E_t}{\text{Maximize}} \int_0^\infty e^{-\delta t} \pi_t \, dt \qquad (3.38)$$

This maximization is subject to the conditions (for $t \geq 0$)

$$\frac{dX_t}{dt} = G(X_t) - q(X_t) X_t E_t \qquad (3.39)$$

$$\frac{dK_t}{dt} = I_t - \gamma K_t \qquad (3.40)$$

$$X_t \geq 0 \qquad (3.41)$$

$$0 \leq E_t \leq K_t \qquad (3.42)$$

$$X_0, K_0 \text{ given} \qquad (3.43)$$

and possibly also to a constraint on the investment rate I_t. This maximization problem is distinctly nontrivial, but solvable (and in an intuitively understandable fashion).

Completely Reversible Investment

We first consider the case in which investment in fishing vessels is completely reversible—the owner can buy or sell vessels, at the going price c_K, without limit. Thus there are *no* constraints on I_t. This case is clearly unrealistic, but quite instructive. [Clark et al. (1979) use the term *completely malleable* to describe the situation in which capital is completely reversible.)

Given that investment is completely reversible and costly, it is evident that the owner will never retain vessels that he is not using. Therefore, we will have

$$E_t \equiv K_t \qquad (3.44)$$

in this case.

Consider the following investment-cost term from Eq. (3.38):

$$\int_0^\infty e^{-\delta t} c_K I_t \, dt = c_K \int_0^\infty e^{-\delta t} \left(\frac{dK_t}{dt} + \gamma K_t \right) dt$$

$$= c_K \int_0^\infty e^{-\delta t} (\delta + \gamma) K_t \, dt - c_K K_0$$

$$= c_K (\delta + \gamma) \int_0^\infty e^{-\delta t} E_t \, dt - c_K K_0$$

where we have used Eq. (3.40), integration by parts, and Eq. (3.44), respectively. Replacing this term in Eq. (3.38), we obtain the problem

$$\underset{E_t}{\text{Maximize}} \int_0^\infty e^{-\delta t}(pqX_t - c_{\text{total}})E_t \, dt \qquad (3.45)$$

where

$$c_{\text{total}} = c + (\delta + \gamma)c_K \qquad (3.46)$$

The investment variable I_t has disappeared (it is given by $I_t = dE_t/dt + \gamma E_t$), and only E_t remains to be determined. Also, the entire costs incurred by the owner are now included in the single term

$$c_{\text{total}}E_t = [c + (\delta + \gamma)c_K]E_t$$

In other words, *when capital is completely reversible, the costs of capital become variable costs.* But this is almost a banality: In essence, the owner merely "rents" vessels as required, paying as rental the costs of interest and depreciation on capital value of the vessels. In fact, completely reversible capital is not really capital at all.

We conclude that, under the assumption of complete reversibility of capital, the sole owner's optimization problem reduces to the problem solved in Chapter 1, in which capital plays no role, but with total variable cost now including both interest and depreciation. For future reference let x^*_{total} denote the optimal equilibrium biomass level for this problem; x^*_{total} is determined by our fundamental rule, Eq. (1.26), but with c replaced by c_{total}. Also let K^*_{total} be the capital (i.e., fleet capacity) required for sustained-yield harvesting at $X = X^*_{\text{total}}$:

$$K^*_{\text{total}} = E^*_{\text{total}} = \frac{G(X^*_{\text{total}})}{q(X^*_{\text{total}})X^*_{\text{total}}} \qquad (3.47)$$

The solution $(X^*_{\text{total}}, K^*_{\text{total}})$ will prove to be important also in the case of irreversible capital.

Irreversible Capital

Capital invested in fishing vessels and gear is doubtlessly more "malleable" than traditional forms of capital such as buildings, railway tracks, and the like. But the owner of a fishing fleet is likely to encounter considerable difficulty in selling off any excess vessels. In common-property fisheries excess capacity is often considered a serious problem, and in most cases the market for vessels *outside* the given fishery is very limited.

For the purpose of providing a sharp contrast to the completely

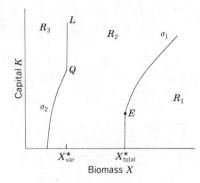

FIGURE 3.7 Feedback control diagram for the case of completely irreversible capital. R_1: Invest at maximum rate; R_2: No investment, fish at full capacity; R_3: No investment or fishing; LQ: Temporary sustained yield; E: Long-term equilibrium. (From Clark et al. 1979.)

reversible case, let us now suppose that capital is completely *irreversible*:

$$I_t \geq 0 \qquad (3.48)$$

This amounts to assuming that there is no resale market whatsoever for excess capacity. Later on we will look at the possibility of resale at a reduced price.

The sole owner's profit-maximization problem is the same as before, except for the additional constraint (3.48). As will soon become apparent, this single change to our model brings in a surprising degree of complexity.

Introduce the notation X_{var}^* for the optimal biomass when *only variable costs* are taken into consideration. Then, since $c < c_{total}$ we have

$$X_{var}^* < X_{total}^* \quad \text{and} \quad K_{var}^* > K_{total}^* \qquad (3.49)$$

The significance of X_{var}^* can be seen as follows.

Suppose for the moment that the owner has plenty of vessels: $K_0 > K_{var}^*$. Since the vessels cannot be sold, fixed costs become irrelevant to the owner's future policy decisions.† Therefore, the optimal biomass is now X_{var}^*. However, the initial capital K_0 depreciates at the rate γ—vessels wear out, are lost at sea, and so on. Eventually we have $K_t < K_{var}^*$. Sustained yield at $X = X_{var}^*$ is no longer possible unless new vessels are brought in. But new vessels involve fixed costs, and the optimal biomass when fixed costs are relevant is X_{total}^*. Thus the existence of irreversible capital seems to give rise to *two* "optimal" equilibrium solutions.

Refer now to Figure 3.7. This is a representation of the positive $X-K$ plane. The plane is divided into three regions R_1, R_2, R_3 by two curves

---

†We ignore such considerations as tax writeoffs and and bankruptcy.

labelled σ_1 and σ_2. These curves, called "switching curves," can be computed numerically; a computer program for this purpose is available from the author.

The sole owner's optimal investment/effort policy can be read off from the Figure. The owner first determines the current values of X_t and K_t. He then chooses his current policy according to the rules:

(X_t, K_t) in	Optimal Policy
R_1	Invest ($I_t = +\infty$), increasing K up to value determined by curve σ_1
R_2	Do not invest ($I_t = 0$); fish at full capacity ($E_t = K_t$)
R_3	Do not invest ($I_t = 0$); do not fish ($E_t = 0$)

There is also a set of positions, on the line QL, where sustained-yield harvesting is optimal, with $X_t = X^*_{\text{var}}$ and with $E_t = E^*_{\text{var}} < K_t$. Because of depreciation, however, these positions are temporary. Finally, a permanent, long-run equilibrium exists at the point $X = X^*_{\text{total}}$, $K = K^*_{\text{total}}$.

This solution is a bit complicated [nearly a year's research was needed to obtain it—see Clark et al. (1979) for rigorous demonstration], but it is quite understandable. For example, note that investment never occurs when $X_t < X^*_{\text{total}}$—this was explained above. Investment does occur if $X_t > X^*_{\text{total}}$, the optimal level of capital being an increasing function (given by σ_1) of the available biomass X_t. The larger the existing resource stock, the greater should be the initial level of investment.

Fishing should not occur ($E_t = 0$) if X_t is below X^*_{var}, although there is also a "premature switching" curve σ_2 (see Section 3.1) which comes into play when capacity K_t is less than K^*_{var}.

When X_t lies between X^*_{var} and X^*_{total} the optimal policy utilizes full fishing capacity ($E_t = K_t$), but no new investment is undertaken. This interval

$$X^*_{\text{var}} < X_t < X^*_{\text{total}} \tag{3.50}$$

is highly significant for management policy; it will be referred to as the *nonmalleability gap*.

Temporary sustained-yield harvesting occurs along the line QL in Figure 3.7, with $X_t = X^*_{\text{var}}$ and with $E_t = E^*_{\text{var}} < K_t$, as discussed earlier. (This phase may never actually occur, depending on parameters and initial conditions.)

Finally, the optimal policy converges to a *unique long-term equilibrium* at point A, with coordinates $(X^*_{\text{total}}, K^*_{\text{total}})$. Note that this is the same equilibrium that we obtained for the completely reversible case. In the very long run, capital must be reversible—it eventually depreciates away. Of course, to remain at E the owner must continually replace his depreciating capital, at the rate $I = \gamma K^*_{total}$.

As an illustration, let us consider the situation of a newly discovered, unexploited fish stock, with no existing capacity. Assume that $X_0 > X^*_{total}$ (otherwise no profits are possible) and $K_0 = 0$. Vessel capacity is purchased at $t = 0$, to the level given by σ_1. This capacity is then fully utilized, leading to fishing down of the biomass. The capital stock K_t slowly depreciates, but is not replaced yet. If X_t falls as low as X^*_{var}, some of the capacity becomes excessive and is removed from the fishery (by assumption it cannot be sold off!). Eventually depreciation takes its toll, and capacity K_t falls below sustainable-yield requirements. The stock begins to recover. When it reaches X^*_{total}, new vessels are purchased, increasing capital to K^*_{total}, and equilibrium prevails.

It may seem amazing that such a complex process is required in order to maximize profits. The complexity arises from the dynamic interaction of two capital stocks (fish and boats); see Kamien and Schwartz (1977).

Antarctic Whaling Example

The model has been fitted to data from the Antarctic baleen whale fishery (Clark and Lamberson 1982). The resulting optimal trajectory is illustrated in Figure 3.8. An initial biomass of 400,000 blue whale units (BWU) is assumed, consisting of the three main species, blue, fin, and sei whales. Capital is measured in terms of "factory units," one factory unit consisting of a whale factory vessel plus associated catcher vessels, assumed total value $20 million. The discount rate is $\delta = 10\%$ per annum. The other parameter values used in the model are listed in Table 3.3.

The optimal initial investment is 13.5 factory units. The cycle to long-run equilibrium takes 31 years, during which the whale stock is reduced to a temporary low of 80,000 BWU. The final long-run equilibrium occurs at 115,000 BWU and requires 1.35 fishing units to harvest the sustainable yield of 4175 BWU per annum. The net present value of this exploitation strategy amounts to $1.30 billion, only 3% of which comes from the eventual sustained-yield phase.

Of course, the actual history of Antarctic whaling did not follow this scenario in any quantitative sense. The time period from the initiation of the pelagic whaling industry in 1926 to its present stage (estimated whale stocks consisted of approximately 75,000 BWU as of 1980) lasted considerably longer than 31 years. Many more years will be required to

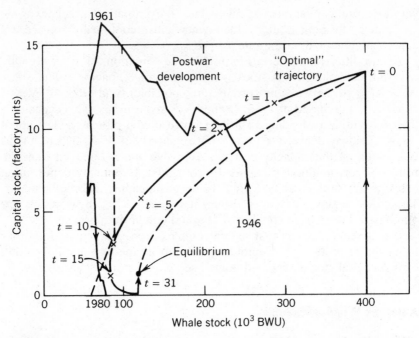

FIGURE 3.8 Profit-maximizing trajectory for a model of Antarctic baleen whales and actual postwar trajectory, 1946–1980. (From Clark and Lamberson 1982, by permission of Butterworths, Guildford, U.K.)

TABLE 3.3 Parameter Values Used in the Antarctic Whale Model

Parameter	Meaning	Value
r	Intrinsic growth rate	.05 per year
X_c	Carrying capacity for Antarctic baleen whales	400,000 BWU
q	Catchability coefficient	.026 per factory unit year
p	Price	$7000 per BWU
c	Variable cost	10^7 per factory unit year
c_K	Cost of capital	2×10^7 per factory unit
γ	Depreciation rate	.15 per year
δ	Discount rate	.10 per year

Source: Clark and Lamberson (1982).

reach equilibrium if the present International Whaling Commission (IWC) moratorium remains in force.

There were actually two distinct investment stages in Antarctic pelagic whaling: a prewar phase initiated by the development of sea-going factory vessels in 1926 and a postwar phase which was needed to replace vessels destroyed during World War II. The postwar expansion reached a peak of 21 factory vessels plus 261 catchers in 1961, but then declined rapidly as whale stocks ran out and increasingly strict quotas were imposed (Fig. 3.8). By 1979 fishing on the stocks of blue, fin, and sei whales had stopped completely, and Minke whales became the only species to be taken legally in the Antarctic.

Although the "optimal" investment strategy differs greatly from the historical record, both have the characteristic of large initial investment levels and an ultimate equilibrium at a relatively low level of capital. In fact, if the current IWC regulations are maintained, the ultimate equilibrium will involve sustained harvesting at 90% of the estimated MSY. According to the model utilized here, this "IWC equilibrium" will have

$$X_{IWC} = 263,000 \text{ BWU}$$
$$H_{IWC} = 4500 \text{ BWU/year}$$
$$K_{IWC} = 0.66 \text{ factory unit}$$
$$\pi_{IWC} = 21.7\$ \text{ million/year}$$

(In fact, each *stock* of Antarctic whales is to be managed separately, so that the simple lumped-variable model used here is somewhat inaccurate.)

We will not attempt to assess in detail the sensitivity of the results described here to the various parameter assumptions, although this is actually an instructive exercise in modelling. But two comments may be in order.

Some whale scientists now think that 5% may be too optimistic a figure for the intrinsic annual growth rate for large whales, and a figure as low as 2% has been suggested (D. Butterworth, personal communication). This parameter change actually affects the optimal trajectory of Figure 3.8 very little, as one might expect from the comment made above that the sustained-yield phase contributes only 3% of total present value. Because of the reduced productivity, the long-run equilibrium position shifts to

$$X^* = 96,000 \text{ BWU}, \qquad K^* = 0.6 \text{ factory unit}$$

and the initial fleet size changes from 13.5 to 12.6 factory units.

Secondly, for various reasons a symmetric logistic growth curve may be unreasonable for whales and should be replaced by a skewed curve with MSY biomass occurring at perhaps 80% of K (May et al. 1979, S. Holt, personal communication). A suitable mathematical form is

$$G(X) = rX \left(1 - \frac{X}{K}\right)^{\alpha} \qquad \text{where } 0 < \alpha \leq 1$$

This function has intrinsic growth rate r and carrying capacity K, but is skewed to the right with

$$X_{\text{MSY}} = \frac{K}{1 + \alpha}$$

which approaches K as $\alpha \to 0$ (e.g., $\alpha = 0.25$ gives $X_{\text{MSY}} = 0.8K$). Also MSY approaches rK as $\alpha \to 0$. For the whale model with $r = .05$ and $K = 400{,}000$, the revised model ($\alpha = .25$) gives MSY = 10,700 BWU/year compared with MSY = 5000 BWU/year for the logistic case.

What effect would this model change have on the profit-maximizing harvest policy? The answer depends sharply on the discount rate. At zero or near zero discounting, the effect is obviously strong. But for $\delta = 10\%$ or larger, the effect is almost negligible. Figure 3.9 depicts the own rate of interest of the Antarctic baleen whale stock, as determined from the skewed model for various values of α. Recall that the own rate of interest is the expression

$$G'(X) - \frac{c'(X)G(X)}{p - c(X)}$$

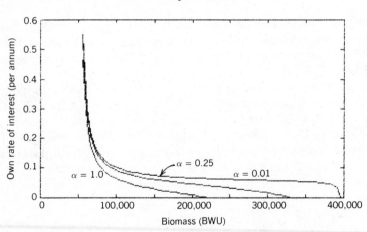

FIGURE 3.9 The own rate of interest for Antarctic baleen whales, using the skewed logistic model, for $\alpha = 1.0$ (nonskewed), $\alpha = .25$, $\alpha = .01$ (highly skewed). The optimal biomass $\hat{X}$ (variable cost) can be read from the figure by locating δ on the vertical axis and reading $\hat{X}$ on the horizontal axis.

which occurs in the golden-rule equation (1.26). The optimal variable-cost biomass $\tilde{X}$ for different interest rates δ can be read directly from these graphs simply by locating δ on the vertical axis and reading off $\tilde{X}$. The effect of α on $\tilde{X}$ is strong for small δ, but weak for large δ, as remarked above. (I have not attempted to redraw Fig. 3.8 for the skewed model, but expect that the cross-effects of α and δ will be similar here.)

Management Implications

Our analysis of irreversible investment yields a number of implications for the management of *transitional* stages in fishery development. First, the development of overcapacity during the initial stages of a new fishery is not necessarily to be avoided completely. However, if expansion is not controlled at this stage, an excessive degree of overcapacity is almost certain to develop (McKelvey 1984; see Table 3.4 and Section 4.2).

More germane is the question of dealing with *existing* excess capacity. The typical situation is one in which (1) biologically based regulation has been imposed, and the stock is being managed at or near the estimated MSY level, but (2) current fishing capacity is grossly excessive. (As noted in Chapter 1, biological management without associated economic controls can have the effect of *increasing* the degree of overexpansion of fishing capacity.)

TABLE 3.4 Actual vs. Optimal Carrying Capacity in Various Fisheries

Fishery	Actual Capacity Built Up	Estimated Optimal Long-Run Capacity[a]
Antarctic whales	26 factory fleets	1–2 factory fleets
British Columbia salmon	6800 vessels	1000–2000 vessels
British Columbia roe herring	1600 licenses	500 licenses
Northern prawns (Australia)	300 vessels	25 vessels
Peruvian anchoveta	1500 vessels	500 vessels[b]
Pribolov seals	108 vessels	0[c]

[a]Estimates based on season length in fully developed (overdeveloped) fishery vs. maximum season length. Optimality of the latter is established in Chapter 4. The initial optimal overcapitalization phase is not considered; it only applies in any significant degree in the case of whales.
[b]Prior to the collapse.
[c]Pribolov seals are best killed on land, as at present.

In this situation it is sometimes recommended that the authorities institute a "buyback" program to remove vessels from the fishery. *The present model* provides no economic justification for such a policy, since the model assumes that

1. Excess capacity has no opportunity cost, that is, no profitable alternative use.

2. Variable costs are proportional to effort E and not to capacity K.

The case in which excess capacity has a positive opportunity cost will be considered below. Regarding variable costs, it may well be the case that annual costs of fishing are positively correlated with capacity, especially when the entire fleet actually participates in the fishery. For example, vessels may have to travel considerable distances to the fishing grounds, and these transportation costs will be determined by the number of vessels participating rather than by the total amount of fishing effort exerted. Alternatively, vessels may rush from one fishing ground to another, following openings of the fishery. In the British Columbia roe herring fishery, for example, vessels have sometimes been transported by helicopter from one fishing area to another (Fraser 1980).

Another problem of practical importance concerns the management of fisheries in which stocks have become "depleted." The simplistic rule is that whenever stocks are depleted the fishery should be closed. But our analysis shows that the concept of "depletion" depends in a rather complicated way on the economic aspects of the fishery. A fish stock may well be "depleted" relative to the *long-run* optimal equilibrium X^*_{total}, but far from depleted in terms of the variable cost optimum X^*_{var}. The model demonstrates that, when capital is irreversible, fishery closure is *not* optimal when the biomass X_t lies between those two levels. The fishery should remain open—but of course any addition of new capacity must be rigorously prevented at this stage.

Further aspects of the problems of optimal and excess capacity will be taken up later.

Alternative Uses

To provide for the possibility that capacity in the fishery may have a positive value for alternative uses, we modify our investment model by removing the constraint $I_t \ge 0$, assuming instead that unwanted capacity can be sold off at a price c_S, where

$$c_S < c_K \tag{3.51}$$

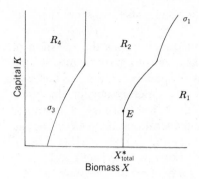

FIGURE 3.10 Feedback control diagram for the case of capital with alternative value. (From Clark et al. 1979).

The sole owner's objective then becomes

$$\text{Maximize} \int_0^\infty e^{-\delta t} \pi_{St}\, dt \tag{3.52}$$

where

$$\pi_{St} = [pq(X_t)X_t - c]E_t - \phi \cdot I \tag{3.53}$$

with

$$\phi = \begin{cases} c_K & \text{if } I > 0 \\ c_S & \text{if } I < 0 \end{cases} \tag{3.54}$$

The solution to this modified investment problem is illustrated in Figure 3.10. An additional region R_4 arises, with the property that disinvestment ($I = -\infty$) occurs whenever the pair (X_t, K_t) falls in this region. Capacity is reduced to a level determined by the curve σ_3. The rest of the diagram is similar to Figure 3.7; in particular, the ultimate equilibrium occurs at the same position (X_{var}^*, K_{var}^*) as in the previous versions of the model. The implications of this diagram are by now fairly obvious, and we shall not discuss them further; see Clark et al. (1979).

3.4 Processing and Marketing

It is certainly not the case, in most fisheries, that fishermen sell their catches directly in the final consumer market. Yet this has tacitly been assumed in all the models discussed so far. In this section, therefore, we introduce a separate processing/marketing sector, assuming for simplicity that the fishermen sell their catches to integrated processing/marketing firms. Under this assumption it can no longer be assumed that the "price" of fish is the same at all levels.

Introducing this additional sector immediately raises the question of relative buying and selling powers of the various participants. Is the processing/marketing firm a monopsonist (meaning a sole purchaser of raw fish)? A monopolist? Do the fishermen merely accept the price offered by the firm, or do they negotiate prices through a union or cooperative? Of course, many other considerations may affect actual contractual arrangements between fishermen and fish buyers. [See Wilson (1980) for a detailed study of fresh fish markets in New England.]

Assume a particular power structure. The type of question we are interested in is whether the fishery can be expected to operate in a socially optimal manner, or if not, what management policies are appropriate.

The Model

We shall continue to use our simple basic model of the fishery (see Clark and Munro 1980):

$$\frac{dX_t}{dt} = G(X_t) - H_t \tag{3.55}$$

$$H_t = qE_tX_t \tag{3.56}$$

(where q may depend on X_t).

To begin with, assume that the processing sector sells its product on a perfectly competitive market, with price p_{proc}. Let $\alpha \leq 1$ denote the *recovery factor*, that is, the weight of processed fish product resulting from 1 unit of raw fish. We shall suppose that α = constant, although in practice α will usually depend on the *size* of fish processed (a parameter not encompassed by the present model—see Section 3.7). Finally, let c_{proc} denote the cost of processing 1 unit of raw fish (same comment!) and let p_{raw} denote the price paid for raw fish. Then

$$\pi_{proc} = (\alpha p_{proc} - c_{proc} - p_{raw})H_t \tag{3.57}$$

represents the net revenue flow to the processing sector. The price p_{raw} is to be considered a parameter; its value will depend on the relative powers of the two sectors (among other things). To simplify notation we rewrite Eq. (3.57) as

$$\pi_{proc} = (p_{net} - p_{raw})H_t \tag{3.58}$$

where $p_{net} = \alpha p_{proc} - c_{proc}$ is simply the "net market price" of 1 unit of raw fish.

Let $C(E)$ denote effort cost. Then net revenue flow to the harvesting sector equals

$$\pi_{\text{harv}} = p_{\text{raw}}H_t - C(E_t) \tag{3.59}$$

In the case that $C(E) = cE$ is linear, bionomic equilibrium will be characterized by zero net revenues in both sectors:

$$\pi_{\text{proc}} = \pi_{\text{harv}} = 0 \tag{3.60}$$

In view of the model equations, we therefore obtain

$$p_{\text{net}} = p_{\text{raw}} = \frac{c}{qX} \tag{3.61}$$

The result is exactly the same as for the original open-access model: bionomic equilibrium is characterized by the condition $p = c/qX$, where p denotes the price received by fishermen (and, here, also the price received by processors, net processing costs).

In the nonlinear case $C''(E) > 0$, bionomic equilibrium in the harvesting sector is characterized by

$$\frac{\partial \pi_{\text{harv}}}{\partial E} = 0 \tag{3.62}$$

while bionomic equilibrium in the processing sector still satisfies

$$\pi_{\text{proc}} = 0 \tag{3.63}$$

(since there are no nonlinearities pertaining to this sector, by assumption). Hence we obtain

$$p_{\text{net}} = p_{\text{raw}} = \frac{c'(E)}{qX} \tag{3.64}$$

which again is the same as the single-sector model [see Eq. (3.32)].

Our first conclusion, then, is that *if* the processing sector is perfectly competitive in the purchasing of fish, then the existence of this sector can be ignored in bioeconomic modelling of the fishery. Similarly, a fully integrated harvesting and processing industry is analytically equivalent to the case of a sole owner. New structures arise only when the two sectors are independent and not both competitive.

Suppose, for example, that the processing sector is monopsonistic, while the harvesting sector remains competitive. We then have from Eq. (3.62), viz. $\partial \pi_{\text{harv}}/\partial E = 0$, that

$$C'(E) = p_{\text{raw}}qX \tag{3.65}$$

Thus, given the price p_{raw} and the current biomass level X_t, the competitive effort level E_t is uniquely determined, as is the catch rate is $H_t = qX_tE_t$ (here we ignore the possibility of temporary disequilibrium in the harvesting sector). This implies in turn that the processing firm can use the price $p_{raw} = p_{raw,t}$ as a *control* to obtain any desired rate of harvest. (Of course, being the sole purchaser of raw fish, by assumption, the processing firm can also *directly* control catches through its purchasing policy.)

From Eqs. (3.58) and (3.65) we now obtain

$$\pi_{proc} = \left[p_{net} - \frac{1}{qX_t} C'(E_t) \right] H_t = \left[p_{net} - \frac{1}{qX_t} C'\left(\frac{H_t}{qX_t}\right) \right] H_t \quad (3.66)$$

The monopsonistic processor's objective is to maximize

$$\int_0^\infty e^{-\delta_{proc}t} \pi_{proc}(X_t, H_t) \, dt \quad (3.67)$$

where δ_{proc} denotes the discount rate used by the firm. As shown in Section 3.2 [see Eq. 3.21)], this problem has an equilibrium solution X^*_{proc}, H^*_{proc} determined by the basic equation

$$G'(X) + \frac{\partial \pi_{proc}/\partial X}{\partial \pi_{proc}/\partial H} = \delta_{proc} \quad (3.68)$$

What is the *social* optimum? Adopting the simplifying assumption that processing and harvesting costs represent true social costs, we see that the social net benefit flow is given by

$$\pi_{soc} = p_{net}H_t - C(E_t)$$

$$= p_{net}H_t - C\left(\frac{H_t}{qX_t}\right) \quad (3.69)$$

The socially optimal equilibrium is therefore determined by the equation

$$G'(X) + \frac{\partial \pi_{soc}/\partial X}{\partial \pi_{soc}/\partial H} = \delta_{soc} \quad (3.70)$$

Two possible sources of divergence exist between the social optimum and the optimum as perceived by a monopsonist: a possible divergence in discount rates, on the one hand, and a concern with total costs vs. marginal costs of harvesting, on the other. The situation is analogous to the divergence between monopoly and socially optimal fishery production discussed in Section 3.2.

Indeed, in the event that the monopsonist also possessed monopoly power, the three sources of divergence would coexist.†

Since by assumption marginal effort costs exceed average costs [except when $C(E)$ is linear], the monopsonist in effect perceives a higher harvesting cost level than does the social manager. Hence the monopsonist tends to *underproduce*, relative to the social optimum, in the sense that

$$X^*_{proc} > X^*_{soc} \quad \text{if} \quad \delta_{proc} = \delta_{soc}$$

As noted previously, however, in the event that $\delta_{proc} > \delta_{soc}$, this divergence is narrowed—or even reversed. (Notice in passing that under the present assumptions—viz. no monopoly power—the social optimum agrees with the integrated sole-owner optimum. Hence the monopsonist is *not* equivalent to a sole owner of the fishery: His perceived fishing costs are marginal rather than total costs; see Anderson 1981.)

The foregoing discussion provides some insight into management policy for fisheries in which the processing/marketing sector is less than perfectly competitive. In general, it follows that restricting one's consideration to the fish harvesting sector alone (as has usually been the case in the literature, at least), may not be appropriate. Management policy (quotas, taxes, subsidies, and the like) may also have to be directed towards the processing sector. In any event, predictions of the impact of management policies will have to take the processing/marketing sector into account. This subject will be taken up again in Chapter 4.

As yet we have not considered the possibility of an organized harvesting sector with bargaining power relative to fish prices. Although this is a fairly common situation (strikes over fish prices are not unusual), it is unfortunately not easily modelled.

It is clear, of course, that a conflict does exist between the interests of fishermen and purchasers of fish. Bargaining between the two sectors will largely be concerned with the division of revenues, as represented by raw fish prices, although there could also be disagreement over management policy. For example, with greater access to capital markets, processing firms might have lower discount rates than fishermen, and if so would tend to favor more conservative harvesting policies. But in practice, harvesting policy is usually fixed by government management authorities, not the fishing industry by itself.

---

†In technical jargon, the divergences not associated with discount rates result from the ability of the monopolist to capture a portion of the "consumers' surplus" and of the monopsonist to capture a portion of the "producers' surplus." These concepts are discussed, in a static framework, by Copes (1972).

In economics, the *theory of games* has been used extensively to model competition between economic agents with conflicting interests. Most published applications, however, have been based on static models; the theory of dynamic games is considered to be one of the most difficult areas of applied mathematics. Some applications of dynamic game theory to fisheries problems will be discussed in Section 4.2.

3.5 Stability

Much of the discussion so far has centered around equilibrium solutions to our models, such as the "bionomic equilibrium" of the open-access fishery, and various "optimal" equilibrium concepts. Obviously, the classical notion of a fixed equilibrium is extremely unrealistic, if taken at face value, in a situation as variable as fish population dynamics. But we do not wish to take up stochastic modelling yet (see Chapter 6); deterministic models can still yield many insights—provided they are not taken too literally.

In the present section we shall discuss briefly the question of *stability* of equilibria of deterministic fishery models. This topic becomes especially interesting (and difficult) for multispecies models, and there is now a large ecological literature on the subject (see May 1975, Holling 1978, also May et al. 1979). In a sense, any fishery is a "multispecies" system—with fish as prey and man as predator. Indeed, some of the ecological principles of predator–prey systems do reappear in the fisheries setting, the main difference being in the time scales of evolution of human technology as compared to the evolution of the behavior of natural predators.

Depensation

A simple setting in which stability questions can arise occurs when the growth function $G(X)$ of the Schaefer model is not convex. A convex function, $G''(X) < 0$, carries the implication that per capita net growth rate, $G(X)/X$, is a decreasing function of population size X. This phenomenon, often referred to as density "compensation," may be attributed to population crowding, depletion of food supply, and so on.

The hypothesis may be formulated that net per capita growth rate of a population might decline also at low population levels. For example, breeding success may decline at very low population levels, for obvious reasons. The same phenomenon may arise when predators remove a fixed

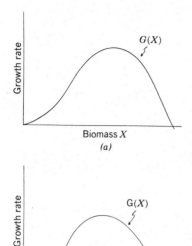

Growth rate

$G(X)$

Biomass X

(a)

Growth rate

$G(X)$

$\underline{X}$ Biomass X

(b)

FIGURE 3.11 Nonconvex growth function: (a) noncritical depensation; (b) critical depensation.

number of prey per unit time, regardless of the magnitude of the prey population. The young fish of anadromous species, such as salmon, may have to "run a gauntlet" of a fixed population of sedentary predators as they pass down spawning streams en route to the sea (Larkin 1966).

Such processes are said to give rise to *depensation* in growth curves —see Figure 3.11. The depensation is termed *critical* if the net growth rate becomes negative at low population levels, that is, for X below some *minimum viable population* level $\underline{X}$. In the latter case, a population is doomed to extinction if it ever falls below the level $\underline{X}$. (More realistic models of extinction would of course be based on probabilistic considerations—see Ludwig 1974.)

Populations exhibiting depensation are capable of providing "surprises" under exploitation. In the case of critical depensation, once the population has been reduced below the minimum viable population level $\underline{X}$, it will not respond to reductions in fishing pressure, but will remain at a low level, approaching extinction. Larkin (1966) suggests that this may be the explanation for the failure of certain salmon stocks to reestablish themselves following severe depletion or accidental reduction of the stock. Any schooling species which is heavily predated, as most are, will tend to have a depensatory recruitment curve, and Clark (1974) suggests that this mechanism could be involved in the dynamics of these species, particularly their common failure to recover. An alternative explanation,

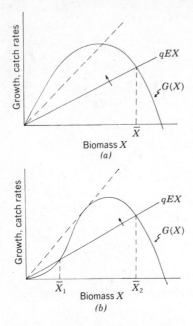

Biomass X
(a)

Biomass X
(b)

FIGURE 3.12   Fishery dynamics under increasing effort E: (*a*) convex case; (*b*) depensation.

in terms of replacement of one species by a competing species, has found favor with biologists—see Murphy (1980). Whichever phenomenon prevails (or if both do), the message is that fish stocks need not have limitless resilience; if sufficiently seriously depleted, stocks may never recover, or only recover following some lucky event leading to an exceptionally large recruitment/stock ratio. Assuming the traditional catch relationship

$$H = qEX \qquad (q \text{ constant})$$

let us consider the response to an increase in effort level E. In the convex case (Fig. 3.12*a*), the equilibrium biomass $\bar{X}$ [determined by $G(\bar{X}) = qE\bar{X}$] depends *continuously* on E. Under depensation, however (Fig. 3.12*b*), there is a *threshold* level of effort E_{thr} with the property that the stable equilibrium $\bar{X}_2$ jumps suddenly to zero as E passes through E_{thr}.

Ecologists have identified a variety of circumstances in which exploited (or otherwise altered) ecosystems may undergo discontinuous shifts to new equilibria (Peterman et al. 1979). Another fishery example, encountered in Chapter 2, arises from a type IV concentration profile. In fact, as shown in Figure 3.13, a Type III profile can have a similar effect.

An important characteristic of all these situations involving discontinuous shifts in equilibria is their *irreversibility*. Figure 3.14 depicts

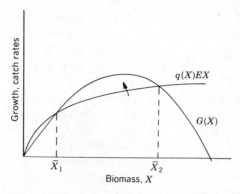

FIGURE 3.13   "Catastrophic" fishery dynamics due to type III concentration profile.

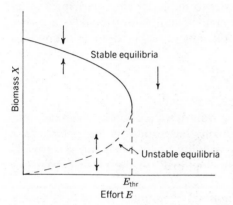

FIGURE 3.14   The equilibrium "manifold" corresponding to Figure 3.12(b) or to Figure 3.13.

the equilibrium "manifold" for the depensation model shown earlier in Figure 3.12(b). This manifold (here merely a curve) is constructed of the equilibrium solutions $\bar{X}_1$ and $\bar{X}_2$, as functions of the control variable E. The lower branch of the manifold consists of unstable equilibria.

Imagine now that, starting with an unexploited stock, effort E_t is gradually increased from a low level. The biomass X_t will first "trace" the upper branch of stable equilibria. But if E_t surpasses the threshold level E_{thr}, the biomass X_t starts to decrease toward 0. But now, even if E_t is decreased to a subthreshold level, the biomass may continue decreasing, since the lower equilibrium X_1 is unstable. In the case of critical depensation, the situation may become completely irreversible, but in other cases a sufficiently drastic reduction in effort would bring about a reversal in biomass decline. [This is an example of "hysteresis" in the "cusp catastrophe"—see Zeeman (1974). Related phenomena can arise in virtually any nonlinear dynamic system, and a huge literature has now developed, identifying possible "catastrophes" in all sorts of situations. For other occurrences in fisheries, see Clark and Mangel (1979) and Peterman et al. (1979).]

The biological reasons why certain fish populations fail to recover after collapses, while others recover successfully, cannot be explained by one simple mathematical model, of course. Our model does, however, suggest the need for especially careful management of the pelagic schooling species. In practice, since depensation is unlikely to be quantitatively predictable, and since entering the depensatory region can lead to irreversible outcomes, it would seem advisable to avoid running down such fish stocks much below the estimated MSY level, regardless of calculations of "optimal" (present-value maximizing) stock levels. Also, the need to monitor fished stocks and to reduce fishing, perhaps drastically, when the stock declines is emphasized. Unfortunately, accurate stock assessment is difficult for patchily distributed populations—see Chapter 6.

Depensation and Extinction

If critical depensation exists for a certain resource stock, that stock may be driven to extinction without actually harvesting the last members of the population. Clearly this is a possibility under open-access exploitation. Would extinction ever appear "optimal" to the profit-maximizing resource owner?

In fact, extinction may appear optimal even under noncritical depensation, and even when our basic marginal productivity formula

$$G'(X^*) = \delta \qquad (3.71)$$

has a positive solution X^*. The simplest way to see this is to compare the present values (PV) of the policy of sustained yield at X^* with the policy of extinction. Assuming $X_0 = X^*$, we have

$$PV_{sust} = \int_0^\infty e^{-\delta t} pG(X^*)\, dt = \frac{pG(X^*)}{\delta}$$

whereas (for immediate extinction)

$$PV_{ext} = pX^*$$

Therefore extinction is the more profitable policy if

$$\frac{G(X^*)}{X^*} < \delta \qquad (3.72)$$

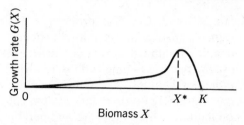

FIGURE 3.15 Strongly depensatory growth curve. If δ is large, the profit-maximizing policy will result in extinction, even though $G'(X) = \delta$ has a positive solution X^*.

It is easy to see that this cannot be the case if $G(X)$ is convex, but it certainly can hold otherwise.

What is the intuitive content of this result? Consider an extreme case of depensation, as shown in Figure 3.15. The sustainable yield $G(X^*)$ is small compared with the required biomass level X^*. Hence harvesting the entire stock may be more profitable than maintaining a sustained yield ad infinitum. This will *not* be the case, obviously, if $\delta = 0$ or if δ is sufficiently small.

The outcome is that (1) the discount rate may be more significant than suggested by our previous theory and (2) in the absence of convexity conditions, "local" optimality formulas may not provide the overall optimum. (The latter fact is very well known to mathematicians but has sometimes been overlooked in the literature on resource economics.)

A great deal more has been written on the topic of optimal extinction (or "exhaustion") of renewable resources, but since most of the issues seem more theoretical than practical, the reader is referred to the literature (Clark 1971, 1973a,b, 1975; Beddington et al. 1975; Clark and Munro 1978; Berck 1979; Cropper et al. 1979; Sinn 1982; Majumdar and Mitra 1983).

3.6   Discrete-Time Models

The view of fish population dynamics provided by the general production model

$$\frac{dX_t}{dt} = G(X_t) - H_t \tag{3.73}$$

is extremely superficial in many ways. In this section we discuss a discrete-time version of the above model, which is not only more realistic (at least for certain species), but also provides a framework for the modelling of certain management techniques, such as annual catch quotas. (A more detailed treatment occurs in Chapter 4.)

The choice between continuous-time and discrete-time models is often portrayed as an either–or choice. This is unfortunate, since models of *mixed* type are often extremely useful (Clark 1976a, chap. 7). Indeed, in the case of fish populations, where breeding and certain other processes occur on a seasonal or annual basis, while some processes (e.g., natural and fishing mortality) occur continuously, the use of mixed-type models seems unavoidable.

(Other classes of mathematical models that have sometimes been used in population dynamics include integral equations and delay-difference equations; such refinements will not be discussed in this book. Age-structured models, which are commonly used in fishery management, will be discussed in the next section.)

The Ricker Model

A discrete-time analog of Eq. (3.73) is

$$X_{k+1} = G(X_k - H_k) \tag{3.74}$$

Here X_{k+1} is interpreted as *recruitment* to the population in year (or cycle) $k + 1$, and $S_k = X_k - H_k$ represents *escapement* the previous year, H_k being the catch. The function $G(S)$ is thus referred to as the *stock recruitment function*. Equation (3.74) is usually referred to as the Ricker stock recruitment model, having been developed by W. E. Ricker (1954) for use in salmon management.

The formulation (3.74) presupposes that recruitment is determined by the total escapement in the previous period. Thus either (1) successive generations do not overlap (as in some species of Pacific salmon) or (2) fish reach sexual maturity by the end of the first period. [For the different species of Pacific salmon, the "period," or life cycle between successive spawnings, is not 1 year, but varies from 2 to 5 years. The Ricker model applies if all fish in the population have the same cycle period; for a more general salmon model, see Chuma (1981).]

A particular form of $G(S)$, developed for use in the Pacific salmon fisheries (Ricker 1954), is

$$G(S) = Se^{a(K-S)} \tag{3.75}$$

Here K represents the natural equilibrium or carrying capacity: $G(K) = K$. The intrinsic rate of growth per period is given by

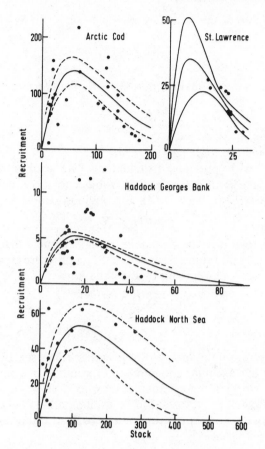

FIGURE 3.16 Ricker curves fitted to data from four gadoid fisheries. (From Cushing and Harris 1973).

$$r = G'(0) - 1 = e^{aK}$$

Figure 3.16 shows the Ricker curve fitted to data from four gadoid fisheries. The fit is obviously pretty tenuous in these examples, which are quite typical of fishery stock recruitment data. We will consider the whole question of "noise" in fishery data (and in fisheries) in Chapter 6.

Optimization Model

Let $\pi_k = \pi(X_k, K_k)$ denote net revenue obtained from harvest H_k extracted from recruitment X_k. (A submodel for π_k is discussed later.)

The present value of the revenues $\pi_0, \pi_1, \ldots$ is given by

$$PV = \sum_{k=0}^{\infty} \alpha^k \pi(X_k, H_k) \qquad (3.76)$$

where α denotes the annual (or periodic) *discount factor*:

$$\alpha = \frac{1}{1+i}$$

where i = annual (periodic) *discount rate*.

In analogy with previous formulations, we assume that the sole owner's objective is to maximize PV, subject to the equation of population dynamics (3.74); harvests H_k are obviously constrained by

$$0 \le H_k \le X_k \qquad (3.77)$$

It can be shown (see the Appendix to this chapter) that this problem has an optimal equilibrium escapement S^*, with corresponding optimal sustained yield H^*, determined by the equations

$$G'(S) \cdot \frac{\partial \pi / \partial R + \partial \pi / \partial H}{\partial \pi / \partial H} = \frac{1}{\alpha} \qquad (3.78)$$

$$H = G(S) - S \qquad (3.79)$$

This is the discrete-time analog of our earlier rule—see Eqs. (3.21) and (3.22)—and it has a similar economic interpretation in terms of marginal productivity. No new bioeconomic insights follow from this reformulation, but it will prove useful later. [Some authors have preferred to develop fishery economics entirely on the basis of discrete-time models; see Levhari et al. (1981).]

Annual Catch Quotas

As an indication of the usefulness of mixed-type models, we next consider briefly the question of total catch quotas, often referred to as TACs—total allowable catches. We discuss fleet capacity and length of the fishing season as they would be expected to develop in an unregulated open-access fishery and in an open-access fishery managed by means of TACs, as well as in the optimal sense. (The present model will be further developed in Chapter 4.)

Our model of fishing during a given season (the continuous-time component of the overall model) is

$$\frac{dx_t}{dt} = - qE_t x_t, \qquad 0 \le t \le T \qquad (3.80)$$

$$x_0 = X, \qquad x_T = S \qquad (3.81)$$

where x_t denotes the biomass at time t during the season, T being the total length of time available for fishing. As before, X and S denote recruitment and escapement, respectively, for the given season. Total season's catch is given by

$$H = \int_0^T qE_t x_t \, dt = -\int_0^T \frac{dx_t}{dt} \, dt = X - S \qquad (3.82)$$

Note that the model ignores any natural mortality (or growth) during the fishing season. Such complications can easily be introduced, but the model becomes somewhat more unwieldy.

Let K now denote total fishing capacity (in SFU) utilized in the given season. Net seasonal revenue will be expressed as

$$\pi = \int_0^T (pqE_t x_t - cE_t) \, dt - c_K K \qquad (3.83)$$

where $c =$ variable cost and $c_K =$ fixed cost. (Note that c_K represents *seasonal* fixed cost; this would include the cost of transporting vessels to the fishing ground and other, similar costs. If vessel capital was completely nonmalleable, costs such as depreciation and interest would be listed as long-run fixed costs rather than as seasonal fixed costs.) Seasonal effort is constrained by capacity:

$$0 \le E_t \le K \qquad (3.84)$$

The expression π in Eq. (3.83) can be written in the form

$$\pi = \int_0^T pH_t \, dt - \int_0^T cE_t \, dt - c_K K$$
$$= pH_{\text{total}} - cE_{\text{total}} - c_K K \qquad (3.85)$$

where H_{total} and E_{total} denote total seasonal catch and total (aggregated) effort, respectively.

The following expression for π will also prove useful:

$$\pi = \int_0^T [p - c(x)]H_t \, dt - c_K K$$
$$= -\int_0^T [p - c(x)]\frac{dx}{dt} \, dt - c_K K$$
$$= \int_S^R [p - c(x)] \, dx - c_K K \qquad (3.86)$$

where

$$c(x) = \frac{c}{qx} \tag{3.87}$$

Consider first the completely unregulated open-access fishery. In a given season, fishing will continue until biomass x_t is reduced to the level $\bar{x}$ at which variable cost equals price:

$$c(\bar{x}) = p \tag{3.88}$$

(unless capacity is too small to permit this—see below). Thus escapement $S = \bar{x}$ in this case, and net revenue in equilibrium is

$$\bar{\pi} = \int_{\bar{x}}^{G(\bar{x})} [p - c(x)] \, dx - c_K \bar{K} = 0$$

Hence bionomic equilibrium capacity K is given by

$$\bar{K} = \frac{1}{c_K} \int_{\bar{x}}^{G(\bar{X})} [p - c(x)] \, dx \tag{3.89}$$

If, however, the cost c_K is large, this value of $\bar{K}$ may be too small to permit fishing to reduce x_t to the level $\bar{x}$. In this case, the bionomic equilibrium values of escapement S and capacity K are determined by the equations

$$G(S) = e^{qKT} S$$

$$\pi = \int_{S}^{G(S)} [p - c(x)] \, dx - c_K K = 0$$

Next, suppose that the fishery is managed by controlling the escapement level S—for example, at the MSY level. This policy will usually be considered necessary only if it results in increased catches:

$$H > \bar{H} \quad \text{and} \quad S > \bar{x} \tag{3.90}$$

Under open-access conditions, equilibrium capacity K is then determined by

$$\pi = \int_{S}^{G(S)} [p - c(x)] \, dx - c_K K = 0$$

or

$$K = \frac{1}{c_K} \int_{S}^{G(S)} [p - c(x)] \, dx \tag{3.91}$$

It follows by comparison with Eq. (3.89) that $K > \bar{K}$. In other words, fishery management based on biological control of the fishery (via escapement control or, equivalently, by enforcement of appropriate TACs) results in an *increase* in fishing capacity relative to the unmanaged fishery.

This rather disquieting prediction is intuitively clear: Preventing depletion results in increased annual catches and revenues. Hence the number of vessels attracted to the fishery also increases. On grounds of economic efficiency, the increase is unjustified since the original capacity K is more than sufficient to handle the increased yield.

The analysis seems to indicate that TAC-based management is economically perverse, at least unless it is accompanied by some form of limited entry. But limited-entry programs have sometimes proved to be less successful economically than anticipated. This important question will be taken up in detail in Chapter 4.

What about employment? Wouldn't the increased catches $H > \bar{H}$ lead automatically to increased employment? Yes—in the processing sector, at least. But whether an increase in the number of *fishermen* is appropriate is less obvious. According to our model, *less* fishing effort is required to capture the increased yields from the rehabilitated stock. If for some reason this is not the case (e.g., type IV concentration profile), then clearly additional fishermen may be required. But if it is true that less effort can catch more fish (following stock recovery), then employing additional fishermen will merely decrease the productivity of labor in the fishery.

However, there is—as always—another dimension to the problem. As long as individual vessels compete for the TAC, there will be a strong incentive to make these vessels as large and as powerful as possible. Individual vessels will become extremely "efficient" at finding and catching fish, but overall *economic* efficiency of the fleet will be *reduced* as a result of *excess* "efficiency" of individual vessels. (Note that the word *efficiency* is being used in two quite different senses here. We should really say that increased vessel *effectiveness* can lead to decreased fleet *efficiency*. The confusion often passes unnoticed in discussion.)

Now the labor requirements of overeffective,—that is, over-capital-intensive—vessels may be much lower than the labor requirements of an efficient fishing fleet. This situation is especially likely to prevail in an underdeveloped country, in which the opportunity costs of labor are close to zero. In such cases, the "small is beautiful" philosophy is likely to concur with economic efficiency in the fishing industry.

How can the incentive for overcapitalization be countered? This is the topic of Chapter 4.

Intra- and Interseasonal Optimization

To complete our present model, we need to consider the overall intra- and interseasonal optimum. The intraseasonal problem is

$$\text{Maximize } \pi \atop (E_t), K \tag{3.92}$$

where π is given by Eq. (3.83), and

$$0 \le E_t \le K$$

$$x_0 = R, \qquad x_T = S$$

Here R and S are to be treated as parameters; they will be determined later by the interseasonal optimization.

From Eq. (3.86) we have

$$\pi = \int_S^R [p - c(x)]\, dx - c_K K$$

so that the optimization in (3.92) is trivial: K should be taken as small as possible, and hence $E_t \equiv K$ for $0 \le t \le T$. This implies that

$$K^* = \frac{1}{qT} \ln \frac{R}{S}$$

and thus

$$\pi^* = \pi^*(R, S) = \int_S^R [p - c_{\text{total}}(x)]\, dx \tag{3.93}$$

where

$$c_{\text{total}}(x) = \frac{c + c_K/T}{qx} \tag{3.94}$$

The interseasonal problem is now

$$\text{Maximize } \sum_{k=0}^{\infty} \alpha^k \pi^*(R_k, S_k) \tag{3.95}$$

where π^* is given by (3.93), and the decision variables are the annual escapement levels $S_0, S_1, S_2, \ldots$. The optimal equilibrium escapement is given by Eq. (3.78), which can be written explicitly here as

$$G'(S) \cdot \frac{P - c_{\text{total}}[G(S)]}{p - c_{\text{total}}(S)} = \frac{1}{\alpha} \tag{3.96}$$

The present problem, moreover, is "myopic" in the sense of Section 3.1 (this is easy to verify—see this chapter's Appendix), which means that the optimal adjustment policy is "bang–bang":

$$H_k^* = \begin{cases} R_k - S^* & \text{if } R_k > S^* \\ 0 & \text{if } R_k \leq S^* \end{cases}$$

We will not discuss the case of irreversible capital here; the results (Charles 1983a) are completely analogous to the continuous-time model of Section 3.3.

3.7 Age-Structured Models

The Leslie matrix model, commonly used in ecology (Emlen 1973) and demography (Keyfitz 1968), can be written as

$$\mathbf{X}_{k+1} = L\mathbf{X}_k \tag{3.97}$$

where $\mathbf{X}_k = (X_k^1, X_k^2, \ldots, X_k^n)$ is a vector whose ith component X_k^i represents the number of individuals in the population of age i in year (period) k. The *Leslie matrix* L is an $n \times n$ matrix of the form

$$L = \begin{bmatrix} f_1 & f_2 & \cdots & f_n \\ \sigma_1 & 0 & \cdots & 0 \\ & & \cdots & \\ 0 & 0 & \sigma_{n-1} & 0 \end{bmatrix} \tag{3.98}$$

Here f_i are *age-specific fecundities* and σ_i are *age-specific mortality rates*, respectively. The maximum age in the population is n years.

If the coefficients f_i, σ_i are constant, the Leslie model is linear and hence predicts exponential long-run growth (or decline) of the population. Such a model is clearly unsuitable for fisheries management, in which nonlinearities resulting from environmental constraints are what determine sustainable yields.

Density-dependent Leslie matrices, in which fecundities f_i or mortality rates σ_i (or both) are assumed to be dependent on the population $\bar{X}$, have received some attention as possible fishery models (Reed 1980, Botsford 1981). These models will not be taken up here, however, since (1) they tend to be complicated and (2) their use depends on estimating several nonlinear, multidimensional relationships, which is seldom feasible from fishery data.

The age-structure models that will be discussed in this section are (1) a "delayed-recruitment" model and (2) the widely used "dynamic pool" model of Beverton and Holt (1957).

A Delayed-Recruitment Model

The following model has been used in the study of fur seal populations (Chapman 1964) and baleen whale populations (Allen 1973). Its mathematical properties have been investigated by Clark (1976b) and Goh and Agnew (1978).

The model equations, analogous to Eq. (3.74) are

$$X_{k+1} = \sigma S_k + G(S_{k-n}) \tag{3.99}$$

$$S_k = X_k - H_k \tag{3.100}$$

Here X_k denotes the adult, breeding population size (biomass) in year k, and S_k denotes escapement of adults in year k. Only adults are harvested. The coefficient σ is the natural survival rate of adults. The function $G(S)$ is the stock-recruitment function; recruitment to the adult population occurs with a *delay* of n years, reflecting the period required to reach sexual maturity. Periods of from 3 to 8 years are common in marine mammals.

The delayed-recruitment model (3.99) can easily be written as a nonlinear Leslie model, but the form given here seems more transparent. Parametrization of the model requires estimation of the survival rate and of the stock-recruitment function $G(S)$. While these estimations are by no means simple, they are clearly less formidable than for the full nonlinear Leslie model.

If no harvesting occurs, Eq. (3.99) becomes

$$X_{k+1} = \sigma X_k + G(X_{k-n}) \tag{3.101}$$

This is a nonlinear difference equation of order $n+1$. The dynamic behavior of such equations can be surprisingly complex and is not yet fully understood from the mathematical point of view (Levin and Goodyear 1980, May 1980, Pounder and Rogers 1980).

A point $\bar{X}$ is an equilibrium solution of Eq. (3.101) if it satisfies

$$(1-\sigma)\bar{X} = G(\bar{X}) \tag{3.102}$$

Whether $\bar{X}$ is stable depends on the values of the three parameters σ, n, and $a = G'(\bar{X})$. The region of stability in the σ,a plane is determined by the following inequalities

$$a_n(\sigma) < a < 1 - \sigma \qquad (0 \le \sigma \le 1) \tag{3.103}$$

Here $a_n(\sigma)$ is a sequence of functions satisfying

$$a_n(\sigma) \uparrow \sigma - 1 \qquad \text{as } n \to +\infty$$

It is easy to verify that $a_0(\sigma) = -1 - \sigma$ and $a_1(\sigma) = -1$; a simple al-

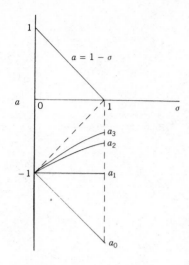

FIGURE 3.17 Stability regions for the delayed-recruitment model. The line $a = 1 - \sigma$ is the upper limit of stability, and the curves $a_n(\sigma)$ are lower limits.

gorithm for calculating $a_n(\sigma)$ numerically for any value of n is given by Clark (1976b). The stability regions for $n \leq 3$ are shown in Figure 3.17. Note that the region of stability shrinks as the delay period increases—this is a typical phenomenon for such models.

The inequalities in (3.103) take on biological significance if the stock-recruitment function is taken in "logistic" form

$$G(X) = rX\left(1 - \frac{X}{K}\right)$$

Writing $M = 1 - \sigma$ (the natural mortality rate of adults), we have, for the equilibrium $\bar{X}$,

$$M\bar{X} = r\bar{X}\left(1 - \frac{\bar{X}}{K}\right)$$

or

$$\bar{X} = \left(1 - \frac{M}{r}\right)K$$

In order to have a positive equilibrium X we must have $r > M$, that is, the "intrinsic recruitment rate" must exceed the natural mortality rate. (If stable, the equilibrium $\bar{X}$, rather than K, represents the natural carrying capacity. If $\bar{X}$ is unstable, it is not clear what would be meant by carrying capacity.)

Since $a = G'(\bar{X}) = 2M - r$, the stability criterion (3.103) now becomes

$$M < r < 3M + b_n(M)$$

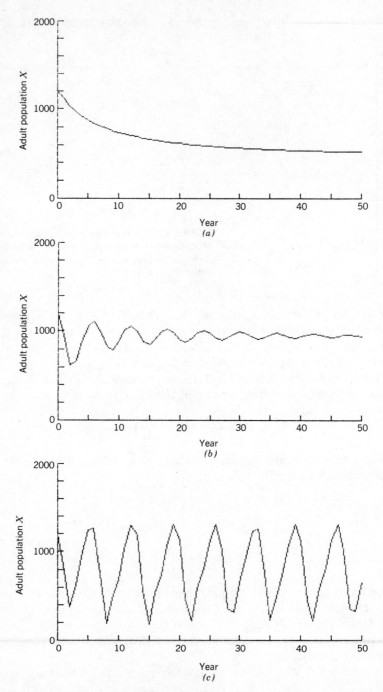

FIGURE 3.18 Simulation of the delay model, Eq. (3.101): (a) stable equilibrium; (b) stable but oscillating equilibrium; (c) unstable equilibrium.

where $b_n(M) \downarrow 0$ as $n \rightarrow +\infty$. Thus stability fails if the intrinsic recruitment rate is too large [see May (1980) for detailed discussion of the biological significance of parameters in delay models].

How does the delay model behave when the equilibrium is unstable? In this case the population sequence $\{X_k\}$ undergoes bounded oscillations, which may be regular or irregular ("chaotic" is the current term). For certain model functions $G(X)$ a negative value of X_k may result, but this anomaly can be avoided by using a growth function $G(X)$ that never becomes negative, such as the Ricker function of Eq. (3.75).

Some simulation results based on the delay model are shown in Figure 3.18; the logistic form was used, with parameter values

$$\sigma = .95, \qquad n = 1, \qquad K = 1000, \qquad r \text{ varying}$$

Since $M = 0.05$, the stability region for r is

$$0.05 < r < 1.10$$

The values $r = 0.10$, 1.0, and 1.5 were used in the simulations shown. The equilibrium X is monotonically stable, oscillatory and stable, and unstable, respectively.

To conclude the discussion of the delay model, let us consider briefly the question of optimal (profit-maximizing) harvest policy. The main question is: What difference does the delay make? To address this question, we compare results for the delay model with results from a corresponding continuous-time nondelay model, using data from the Antarctic fin whale fishery.

The optimization problem for the delay model is (see Clark 1976b for more details):

$$\text{Maximize} \sum_{k=0}^{\infty} \alpha^k \pi(X_k, H_k) \tag{3.104}$$

subject to

$$X_{k+1} = \sigma(X_k - H_k) + G(X_{k-n} - H_{k-n}) \tag{3.105}$$

$$0 \leq H_k \leq X_k \tag{3.106}$$

where [cf. Eq. (3.86)]

$$\pi(X, H) = \int_{X-H}^{X} [p - c(x)] \, dx \tag{3.107}$$

(fixed costs are ignored, for simplicity).

It can easily be shown, using the method of Lagrange multipliers (see

TABLE 3.5 Optimal Escapement Levels for Antarctic Fin Whales: (a) Delayed-Recruitment Model; (b) Logistic Model

Discount Rate (per annum)	Optimal Escapement (number of whales)	
	(a) Delay	(b) Logistic
0	219,302 whales	220,000 whales
.01	195,249	190,229
.03	148,752	140,008
.05	111,752	105,342
.10	70,335	68,811
.15	58,183	57,682
.20	52,958	52,777

Source: Clark (1976b).

the Appendix) that this problem has an optimal equilibrium escapement S^*, determined by the equation

$$[\sigma + \alpha^n G'(S)] \frac{p - c[\sigma s + G(S)]}{p - c(S)} = \frac{1}{\alpha} \qquad (3.108)$$

This is directly analogous to the result for the nondelay model—see Eq. (3.96). The optimal escapement levels for Antarctic fin whales, as calculated from this formula, are shown in the second column of Table 3.5. Parameter values employed were

$$r = 0.12 \text{ per annum}$$
$$\sigma = .96 \text{ per annum}$$
$$K = 600,000 \text{ whales}$$
$$n = 5 \text{ years}$$
$$\frac{c}{pq} = 40,000 \text{ whales}$$

The natural equilibrium is $\bar{X} = 400,000$ whales, and this is stable. The value $c/pq = 40,000$ whales represents the "zero profit" or bionomic equilibrium stock level.

In setting up the continuous-time analog to the delay model, the question arises as to the appropriate value for the intrinsic growth rate r_c. A first guess might be to set $r_c = r - M$, the difference between the "intrinsic recruitment rate" and the natural mortality rate. For fin whales this gives $r_c = .12 - .04 = 8\%$ per annum. Because of the delay, however, this value is too high. (The model already takes care of juvenile mortality.) The correct value, $r_c = 5.23\%$ per annum, must be derived via linearization of the delay model about its unstable equilibrium $X = 0$ (see Clark 1976b).

The optimal biomass levels for the nondelay model, as computed from Eq. (1.26), are given in the third column of Table 3.5. [The discrete-time nondelay model, Eq. (3.96), gives virtually identical results.] Parameter values were

$$r_c = 0.0523 \text{ per annum}$$
$$K = 400,000 \text{ whales}$$
$$\frac{c}{pq} = 40,000 \text{ whales}$$

The two models give very similar results, never differing by more than 3%, a figure well below the level of accuracy in estimating whale stocks and also much lower than the effect of minor changes in the discount rate. Once an appropriate correction in the intrinsic growth rate has been made, the delay phenomenon can therefore probably be ignored in studies of management policy.

Dynamic Pool Models

One of the most commonly used models in fishery management is the "dynamic pool" model developed (for North Sea demersal fisheries) by Beverton and Holt (1957). This model has several attractive properties. First, its parameters have a clear biological interpretation and can be estimated from fishery data. The model appears to allow one to ignore the tricky issue of stock-recruitment relationships (but we will have more to say on this). Finally, the model permits consideration of the effects of both fishing mortality and age of first capture.

The original model of Beverton and Holt will be presented with slightly simplified notation. Let $N_k(t)$ denote the number of fish belonging to the kth cohort and alive at time t. Here t denotes time as measured by calendar date; the kth cohort consists of fish that recruit to the fishery in year k, where by *recruit* we mean "become available to the fishery." The age of a fish in the kth cohort, as measured from the date of recruitment, is equal to $t - k$. We have

$$N_k(t) = 0 \qquad \text{for } t < k$$

$$N_k(k) = R_k \qquad\qquad\qquad (3.109)$$

where R_k denotes recruitment to the kth cohort. Recruitment is usually assumed to occur instantaneously, although this obviously somewhat unrealistic.

Denote by Z_{kt} the total mortality rate for the kth cohort; we assume that

$$Z_{kt} = M + F_{kt} \qquad (3.110)$$

where the natural mortality rate M is assumed constant and where F_{kt} represents fishing mortality. We then have

$$\frac{dN_k(t)}{dt} = -(M + F_{kt})N_k(t) \qquad (3.111)$$

Hence

$$N_k(t) = R_k \exp\left[-\int_k^t (M + F_{kt})\, dt\right], \qquad t \geq k \qquad (3.112)$$

Now let $w(a)$ denote the (average) weight (kg) of a fish of age a. A convenient parametric form with some biological justification is the von Bertalanffy curve:

$$w(a) = w_\infty(1 - e^{-k(a-a_0)})^3 \qquad (3.113)$$

in which weight approaches an asymptotic value w_∞. The expression

$$B_k(t) = N_k(t)w(t - k), \qquad t \geq k \qquad (3.114)$$

represents the *total biomass* of the kth cohort at time t. The *natural cohort biomass* curve arises when $F_{kt} \equiv 0$; it has the equation

$$B_{k0}(t) = R_k e^{-M(t-k)} w(t - k) \qquad (3.115)$$

An example was given in Chapter 1—see Figure 1.10. The natural cohort biomass reaches a maximum at the age $\hat{a}$ (time $\hat{t} = k + \hat{a}$) for which

$$\frac{w'(\hat{a})}{w(\hat{a})} = M \qquad (3.116)$$

Clearly, the maximum possible yield from a given cohort would be achieved if the entire cohort could be harvested at the age $\hat{a}$, but this would require the application of infinite fishing mortality at the instant $\hat{t} = k + \hat{a}$.

Yield Isopleths

In practice, fishing effort is constrained to a finite maximum value: $F_{kt} \leq F^{\max}$. Suppose now that a constant fishing mortality F is applied to a given cohort, starting with fish of age a_c, the age of first capture. (For example, a_c might be determined by the mesh size of nets used in the fishery.) We can then calculate the total yield taken from the cohort, as a function of F and a_c:

$$Y(F, a_c) = \int_{k+a_c}^{\infty} FN_k(t)w(t-k) \, dt \tag{3.117}$$

Using Eq. (3.112), we can write this, after some algebraic simplification, as

$$Y(F, a_c) = FR_k e^{Fa_c} \int_{a_c}^{\infty} e^{-(M+F)u} w(u) \, du \tag{3.118}$$

(For simplicity we allow the age of fish to extend to $+\infty$; this causes only minor inaccuracies and is easily replaced by a more realistic assumption if necessary.) If $w(u)$ is given explicitly by Eq. (3.113), the expression for $Y(F, a_c)$ can also be calculated explicitly (Beverton and Holt 1957, p. 36).

The level contours $Y(F, a_c) = $ constant are called *yield isopleths*; see Figure 3.19. Note that $Y(F, a_c)$ approaches a maximum value as $F \to +\infty$ and $a_c \to \hat{a}$; this is the theoretical yield-maximizing solution noted above.

The Beverton–Holt model has been used as the basis of various concepts of "optimal" fishing; most of these have been independent of economic considerations; all have ignored discounting—which often has a strong influence.

For any fixed level of fishing mortality F, there is a corresponding age of first capture $a_c < \hat{a}$ such that yield $Y(F, a_c)$ is a maximum. The resulting maximum yield curve is called the *eumetric yield*:

$$Y_{\text{eumetric}}(F) = \max_{a_c} Y(F, a_c) \tag{3.119}$$

On the yield–isopleth diagram (Fig. 3.19), eumetric points are the points of vertical tangency of the isopleths. Some authors have proposed that only eumetric points be considered as possible optima.

A dual concept, *cacometric yield*, is defined by maximizing with respect to fishing mortality, keeping a_c fixed:

$$Y_{\text{cacometric}}(a_c) = \max_{F} Y(F, a_c) \tag{3.120}$$

This obviously gives different combinations of F and a_c than for the case of eumetric yield (Fig. 3.19), and it is not clear which of the two concepts is to be preferred. The difficulty arises, of course, from failure to stipulate

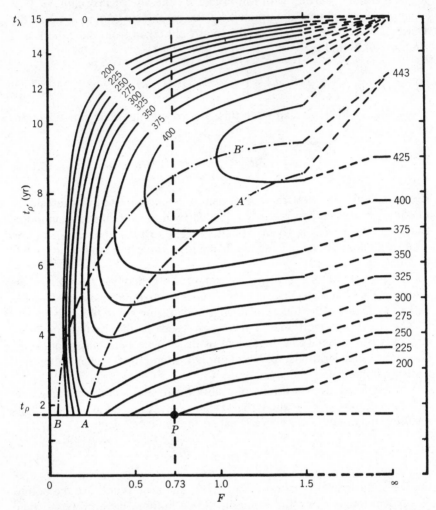

FIGURE 3.19 Yield–isopleth diagram for North Sea plaice. (From Beverton and Holt 1957, p. 318. Reproduced with the permission of the Controller of Her Britannic Majesty's Stationary Office.)

any specific optimization objective—maximum *yield* being operationally infeasible.

The Case of a Single Cohort

Most fisheries exploit several cohorts simultaneously—a given cohort takes several years to "pass through" the fishery. This leads to additional complications in determining optimal fishing strategy, particularly where (as is usually the case) recruitment is subject to large fluctuations.

For the moment, let us ignore these complications and consider the problem of optimizing economic yield from a single cohort. To simplify notation, now let $t = 0$ be the time at which the cohort recruits to the fishery. With the usual simplifying assumptions, the problem can be expressed as:

$$\underset{(E_t)}{\text{Maximize}} \int_0^\infty e^{-\delta t}(pqN_tw_t - c)E_t \, dt \qquad (3.121)$$

subject to

$$\frac{dN_t}{dt} = -(M + qE_t)N_t, \qquad t \geq 0; \qquad N_0 = R \qquad (3.122)$$

$$E_t \geq 0 \qquad (3.123)$$

This problem is myopic in the sense of Section 3.1 and can be expressed in the form studied there. Let

$$\pi_t = (pqw_tN_t - c)E_t = \phi_t(N_t)H_t \qquad (3.124)$$

where $H_t = qN_tE_t$ is the catch rate (in numbers) and

$$\phi_t = pw_t - \frac{c}{qN_t} \qquad (3.125)$$

which represents the net revenue per unit catch. The singular solution N_t^* is then given by Eq. (3.9), which reduces to (Clark 1976a, p. 277)

$$N_t^* = \frac{c\delta}{pqw_t(M + \delta - \dot{w}_t/w_t)} \qquad (3.126)$$

The optimal harvest policy is shown, in terms of cohort biomass, in Figure 3.20. Fishing occurs during a period

$$t_\delta \leq t \leq \hat{t} \qquad (3.127)$$

where $\hat{t}$ is the time (age) of maximum natural cohort biomass and where t_δ is defined by

$$\frac{w'(t_\delta)}{w(t_\delta)} = M + \delta \qquad (3.128)$$

Thus, the larger the discount rate, the more are catches shifted toward the beginning of the cohort's life span and the smaller the total catch from the cohort. The open-access "equilibrium" maintains the biomass at the zero-profit level $\bar{B} = c/pq$ as long as possible (i.e., until $t = \hat{t}$); this coincides with the present-value maximizing policy in the limit as $\delta \to \infty$.

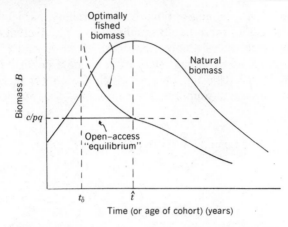

FIGURE 3.20 Single-cohort model. (See text for details.)

These predictions are analogous to results derived from the models discussed earlier, but the mechanism is now somewhat different. It is customary to use the term *recruitment overfishing* to refer to the phenomenon wherein depleted stocks result in a loss of reproduction and subsequent recruitment.

Cohort fisheries may be subject also to *growth overfishing*, by which is meant the removal of too many small fish from each cohort, so that the total yield from the cohort is less than the optimum. Our analysis of the single-cohort model shows that growth overfishing will be most extreme in the unregulated open-access fishery, which will capture fish as soon as they are large enough to be commercially valuable. An excessive level of fishing effort will be applied, and overly nonselective fishing gear will also be employed. Management will thus be directed to the regulation of both gear characteristics and effort levels. Both of these measures may encounter enforcement difficulties (cheaters prosper).

An additional source of conflict can arise if the management authorities base their calculations of optimum fishing on an implicit zero rate of discounting (as appears universally to be the case). Discounting at normal rates can in fact have a strong effect on optimal harvest policy, in terms of age of first capture (and hence the average size of fish in the catch), and total fishing mortality. For example, using Beverton and Holt's (1957) data for North Sea plaice, Clark et al. (1973) found that a 20% discount rate would reduce the optimal age of first capture from 13.5 to 7 years and reduce average yield per recruit by approximately 35% (in comparison with the theoretical, but infeasible, maximum yield). Computer modelling of the Pacific halibut fishery (Bledsoe et al. 1974) has shown similar sensitivity of profit-maximizing harvest policy to the discount rate.

Multiple Cohorts

Observe that the yield isopleths (Fig. 3.19) are independent of recruitment R. This has led to the idea that *yield per recruit* (Y/R) should be the main concern of fishery policy, and much of the literature has taken this assumption for granted. It is a convenient assumption, since Y/R is the same for all cohorts, and hence the fact that cohorts of different size are intermingled in the catch can be ignored.

A currently popular rule of thumb, based on the Y/R idea, is the so-called $F_{0.1}$ management objective (Gulland and Boerema 1973), which can be described as follows. Assume that the age of first capture $\bar{a}_c$ is fixed. Consider yield per recruit as a function of F:

$$\frac{Y}{R} = \frac{Y(\bar{a}_c, F)}{R}$$

(see Fig. 3.21). Let b denote the slope of this curve at the origin: $b = [\partial(Y/R)/\partial F]_{F=0}$. Then $F_{0.1}$ is the value of F for which the slope equals $0.1 \times b$:

$$\frac{\partial(Y/R)}{\partial F}\Bigg|_{F=F_{0.1}} = (0.1) \times \frac{\partial(Y/R)}{\partial F}\Bigg|_{F=0}$$

Although obviously completely ad hoc, the $F_{0.1}$ rule has been adopted as a management objective in several fisheries. Originally, the rule seems to have been based on some notion of "marginal return" to fishing effort (Gulland 1968). Its most attractive feature, however, is probably its built-in conservatism. Since $F_{0.1} < F_{max}$ (the yield-maximizing F), the $F_{0.1}$ level of fishing mortality should be relatively "safe," even if F_{max} has

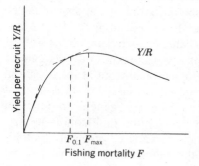

FIGURE 3.21 The $F_{0.1}$ rule: At $F = F_{0.1}$ the slope of the yield–per recruit curve equals 0.1 times its slope at $F = 0$.

been overestimated. Also, yield-per-recruit analysis tacitly assumes that recruitment is exogenous (i.e., no stock–recruitment relationship exists); no one believes that this can be true when stocks are severely depleted, and the $F_{0.1}$ rule again provides a safety margin. Of course, $F_{0.1}$ is finite even when $F_{max} = +\infty$, as is the case for $\hat{a}_c \geq \hat{a}$.

The main defect of rules of thumb such as $F_{0.1}$ is that they largely ignore economic considerations. For example, the result of applying the $F_{0.1}$ rule to an open-access fishery is predictable on the basis of our general theory. Following an adjustment phase (which may be traumatic if the stock is initially depleted), fishing capacity will expand to a greater level than the unregulated equilibrium, and the fishing season will have to be correspondingly reduced. Similar effects will arise from gear restrictions affecting the age of first capture. While such measures may succeed in protecting fish stocks and maintaining yield, they can have economically perverse consequences unless accompanied by regulations that are specifically directed toward economic objectives.

It is clear from Figure 3.20 that the optimal total amount of effort applied to a given cohort depends on the size of the cohort and on the cost/price ratio. Thus the $F_{0.1}$ rule (or any such rule requiring a fixed level of effort) may be particularly inappropriate for fisheries with strong fluctuations in recruitment or in price. However, in the presence of excess capacity, where the fishery would normally impose levels of fishing mortality greater than $F_{0.1}$ on all cohorts, the rule may be necessary to prevent overfishing. In a sense, noneconomic management policy is a vicious circle—the policy results in ever greater overcapacity, which makes the policy all the more necessary, all the more difficult to enforce, and so on.

We should conclude this section, no doubt, with a description of the economically optimal harvest policy for the multicohort fishery in terms of present value maximization. We shall not do so—for two reasons. First, multidimensional dynamic optimization problems are notoriously difficult to solve, and no complete solution has yet been obtained for the case of a multicohort fishery model (some partial results appear in Clark 1976a, sec. 8.4; Botsford 1981; Deriso 1980; Feichtinger 1982; Hannesson 1975; Getz 1980). Secondly, the value of such detailed theoretical results seems very limited; few additional bioeconomic insights can be anticipated, and practical implementation is probably out of the question.

It is clear from the analysis of the single-cohort model that the general bioeconomic principles derived for the Schaefer general production model remain valid—at least qualitatively—for the multicohort fishery. Open-access overexploitation, economic "perversity" of purely biological management policy, sensitivity of the sole owner's optimal harvest policy to the discount rate—all remain valid, as do the considerations of demand and supply, processing and marketing sectors, search and capture techniques, and so on. One difficult and unsolved problem is how to adjust quotas to stock fluctuations in an economically optimal manner.

3.8   Summary and Perspective

In this chapter we have shown that the simple bioeconomic models described in Chapter 1 can be adapted to study a variety of issues and phenomena in fishery management. Time variations in biological and economic parameters, demand schedules and cost nonlinearities, fixed costs and investment, marketing and processing activities, seasonal effects, and age structure have all been discussed. Dynamic bioeconomic modelling is seen as a powerful approach to the understanding of complex issues involved in fishery management. In the following chapter this theory will form the basis for the study of regulatory policy.

Many important dimensions of complexity have still been ignored, however. Our economic analysis has been restricted to the two extremes of sole ownership and pure competition (open access). In reality, intermediate cases are most common—especially in fisheries where entry is controlled by licensing. But the theory of economic systems with finitely many agents is exceedingly difficult and, in any dynamical sense, poorly developed. A very simple fishery model along these lines will be presented in Section 4.2.

On the biological side, no recognition has been given to the complexities of the marine ecosystem. Current management practice seems to be based almost exclusively on single-species models. When applied to a multispecies system, this approach may attempt to maximize yield from each of several interdependent stocks—a logical and practical impossibility. While multispecies models can readily be formulated in the abstract, they present two problems. First, they are difficult to analyze, and results may be model-dependent. More significant, current knowledge of marine ecosystems is too fragmentary to support the development of reliably predictive, usable multispecies models. This problem of "structural uncertainty" is not close to being resolved, and present management of multispecies fisheries seems severely hampered by the lack of any overall paradigm (May et al. 1979; May 1984). Some aspects of the problem are discussed in later chapters.

Deterministic models are of course quite unrealistic even at the single-species level. Models involving random fluctuations and uncertainty are currently being studied by many workers and will be reported on in Chapter 6.

Appendix

In this Appendix we apply some standard techniques of dynamic optimization to the various models discussed in Chapter 3. The principal method employed will be the *maximum principle* for optimal control problems (in continuous time). No attempt will be made to derive or to explain the maximum principle or even to delineate the precise conditions under which it can be rigorously employed. A somewhat more detailed exposition can be found in Clark (1976a, chap. 4); a useful introduction to dynamic optimization problems in economics is given by Kamien and Schwartz (1981).

The reader should perhaps be advised that the maximum principle is a subtle tool, and it has often been seriously misused—some would say abused. For optimization problems that are more complicated than the simple types used in this book, the problem of determining optimal controls is usually formidable.

The Maximum Principle

Consider the following "optimal control" problem:

$$\text{Maximize} \int_0^\infty \alpha(t)\pi_t(\mathbf{X}_t, \mathbf{U}_t)\, dt \qquad (3.129)$$

subject to the dynamic equation

$$\frac{d\mathbf{X}_t}{dt} = \mathbf{G}_t(\mathbf{X}_t, \mathbf{U}_t), \qquad \mathbf{X}_0 \text{ given} \qquad (3.130)$$

and to control constraints

$$\mathbf{U}_t \in A \qquad (3.131)$$

Here $\mathbf{X}_t$ is an n vector, representing the *state* of a certain system at time t, and $\mathbf{U}_t$ is an m vector, representing the *control variable*. The set A is a fixed control set. Equation (3.130) is usually called the *state equation*, and the integral in Eq. (3.129) is the *objective functional*.

The formulation given here is appropriate for economic problems, but more general formulations are used in engineering and other fields. In particular, we assume an *infinite time horizon* in the objective, Eq.

(3.129), which also involves the usual *discount factor* $\alpha(t)$—see Eq. (3.5).

The maximum principle consists of a number of *necessary conditions*, which must be satisfied if $\mathbf{U}_t$ is an optimal control; by an *optimal control* we mean a control variable $\mathbf{U}_t$ that provides the desired maximum for the stated problem.

In using the maximum principle, one normally examines the necessary conditions in the hope of discovering a control $\mathbf{U}_t$ (or perhaps several such) satisfying the conditions. The step from this operation to concluding that in fact $\mathbf{U}_t$ is the desired optimum is usually a difficult one, and it is rather often omitted, not infrequently with hopeful reference to largely fictitious—or irrelevant—"sufficiency conditions." We will take this tradition to an extreme here and simply ignore the problem of rigorous proof of optimality.

The maximum principle, then, can be formulated in terms of the expression

$$\mathcal{H}(\mathbf{X}_t, t, \mathbf{U}_t, \boldsymbol{\lambda}_t) = \pi_t(\mathbf{X}_t, \mathbf{U}_t) + \boldsymbol{\lambda}_t \cdot \mathbf{G}_t(\mathbf{X}_t, \mathbf{U}_t) \qquad (3.132)$$

called the (current-value) *Hamiltonian* of the problem. Here $\boldsymbol{\lambda}_t$ is an ancillary n vector, referred to as the *adjoint variable* (or, in economic problems, as the "shadow price" of the stock $\mathbf{X}_t$); the second term on the right side of Eq. (3.132) is the usual dot (inner) product of the two n vectors $\boldsymbol{\lambda}_t$ and $\mathbf{G}_t(\ldots)$. Most of the applications discussed here will have one-dimensional state (and adjoint) variables, so this is the ordinary algebraic product.

The maximum principle asserts that, if $\mathbf{U}_t$ is an optimal control, then there exists an adjoint variable $\boldsymbol{\lambda}_t$ satisfying the conditions

$$\frac{d\boldsymbol{\lambda}_t}{dt} = \delta_t \boldsymbol{\lambda}_t - \nabla_X \mathcal{H} \qquad (3.133)$$

where $\nabla_X = (\partial/\partial X^1, \ldots, \partial/\partial X^n)$ and

$$\mathcal{H}(\mathbf{X}_t, t, \mathbf{U}_t, \boldsymbol{\lambda}_t) = \max_{\mathbf{U} \in A} \mathcal{H}(\mathbf{X}_t, t, \mathbf{U}, \boldsymbol{\lambda}_t) \qquad (3.134)$$

for all $t \geq 0$. Equation (3.133) is usually called the *adjoint equation*. (Certain other necessary conditions, including the so-called transversality conditions, will not be discussed here.) It is worth noting that the equations (3.130) and (3.133) constitute a system of $2n$ first-order differential equations for the components of the vectors $\mathbf{X}_t$ and $\boldsymbol{\lambda}_t$, with an additional unknown m vector $\mathbf{U}_t$. Equation (3.134), however, yields an additional m conditions; in principle; $\mathbf{U}_t$ can be eliminated by first solving Eq. (3.134). Thus, as in the classical calculus of variations, the original optimization problem is transformed into a system of ordinary nonlinear differential equations.

As a first application, we consider the nonlinear problem of Section 3.2—see Eqs. (3.18) et seq.:

$$\text{Maximize} \int_0^\infty e^{-\delta t}\pi(X_t, H_t)\, dt \tag{3.135}$$

subject to

$$\frac{dX_t}{dt} = G(X_t) - H_t \tag{3.136}$$

The Hamiltonian is

$$\mathscr{H} = \pi(X_t, H_t) + \lambda_t[G(X_t) - H_t]$$

Equation (3.134) implies that $\partial \mathscr{H}/\partial H = 0$, or

$$\lambda_t = \frac{\partial \pi}{\partial H_t}$$

(here we ignore constraints on H_t). Then Eq. (3.133) becomes

$$\frac{d\lambda_t}{dt} = \delta\lambda_t - \frac{\partial \mathscr{H}}{\partial X_t}$$

$$= \delta\lambda_t - \frac{\partial \pi}{\partial X_t} - \lambda G'(X_t)$$

$$= [\delta - G'(X_t)]\frac{\partial \pi}{\partial H_t} - \frac{\partial \pi}{\partial X_t} \tag{3.137}$$

An *equilibrium* solution occurs when $d\lambda_t/dt = 0$ and $dX_t/dt = 0$, that is, provided that X, H satisfy

$$G'(X) + \frac{\partial \pi/\partial X}{\partial \pi/\partial H} = \delta \tag{3.138}$$

$$H = G(X)$$

These are precisely the equations (3.21) and (3.22) quoted in the text.

The optimal adjustment policy can also be calculated from the maximum principle: Eqs. (3.136) and (3.137) are just a system of two differential equations for the two unknown functions X_t, λ_t. While not amenable to analytic solution, these equations can be solved numerically. An example (and further discussion) is given by Clark (1976a, p. 165).

Observe that the possibility of discussing the solution in terms of "adjustment toward an optimal equilibrium" depends upon the existence of an equilibrium solution (X, H) to the system (3.136), (3.137) of differential equations. This system does in fact have an equilibrium solution, because it is autonomous (no explicit time dependence in the equation). If the parameters of the underlying model had been time-dependent, no such equilibrium would have emerged, and only the numerical solution method could have been employed.

It is fortunate, therefore, that nonautonomous models can also be analyzed, provided they have an appropriate linearity property. To illustrate this possibility, we now consider the problem of Section 3.1:

$$\text{Maximize} \int_0^\infty \alpha(t)[p_t q_t(X_t) X_t - c_t] E_t \, dt \tag{3.139}$$

subject to

$$\frac{dX_t}{dt} = G_t(X_t) - q_t(X_t) E_t X_t \tag{3.140}$$

and

$$0 \le E_t \le E^{\max} \tag{3.141}$$

We now have

$$\mathcal{H} = (p_t q_t X_t - c_t) E_t + \lambda_t [G_t(X_t) - q_t E_t X_t]$$
$$= \sigma_t E_t + \lambda_t G_t(X_t) \tag{3.142}$$

where

$$\sigma_t = (p_t - \lambda_t) q_t X_t - c_t \tag{3.143}$$

Now consider condition (3.134) of the maximum principle: E_t must maximize the Hamiltonian $\mathcal{H}$ over the allowable range $0 \le E_t \le E^{\max}$. This obviously implies that

$$E_t = \begin{cases} 0 & \text{if } \sigma_t < 0 \\ E_{\max} & \text{if } \sigma_t > 0 \end{cases} \tag{3.144}$$

and this explains the term *switching function* for σ_t. The most important case, however, arises when $\sigma_t \equiv 0$; this leads to the singular, or myopic, solution, as follows.

If $\sigma_t \equiv 0$, we have

$$\lambda_t = p_t - \frac{c_t}{q_t X_t} = \phi_t(X_t)$$

Hence

$$\frac{d\lambda_t}{dt} = \frac{\partial \phi_t}{\partial t} + \frac{\partial \phi_t}{\partial X_t} \frac{dX_t}{dt}$$

By the adjoint equation (3.133),

$$\frac{d\lambda_t}{dt} = \delta_t \lambda_t - (p_t - \lambda_t) E_t \cdot \frac{d}{dX_t} (q_t X_t) - \lambda_t \frac{\partial G_t}{\partial X_t}$$

Equating these two expressions for $d\lambda_t/dt$, we obtain (after some algebraic manipulations)

$$\frac{\partial G_t}{\partial X_t} + \frac{G_t \partial \phi_t/\partial X_t}{\phi_t} = \delta_t - \frac{\partial \phi_t/\partial t}{\phi_t} \tag{3.145}$$

and this is the same as Eq. (3.9) for the singular solution X_t^*.

The upshot of these calculations is that either we are on the singular path ($\sigma_t \equiv 0$) or E_t must be either 0 or $E^{\max}$. Additional calculations are required to determine whether $E_t = 0$ or $E^{\max}$ when off the singular path ("blocked intervals" may complicate the issue), but we will not pursue the problem further here.

The model of irreversible capital discussed in Section 3.3 is also a linear control problem (autonomous but two-dimensional). The maximum principle can be used to show that the problem has two singular solutions X_{var}^* and X_{total}^*. These turn out to be important for the overall solution, but—as noted in the text—the adjustment dynamics are somewhat complex.

Finally, we discuss briefly the optimization problem for discrete-time models. The equation for optimal equilibrium can be obtained rapidly from a straightforward application of the method of Lagrange multipliers. The method is easily applied to the delayed recruitment model of Section 3.6; the nondelay model is then a special case.

The problem [see Eqs. (3.104) et seq.] is

$$\text{Maximize} \sum_{k=0}^{\infty} \alpha^k \pi(X_k, H_k) \tag{3.146}$$

subject to

$$X_{k+1} = \sigma(X_k - H_k) + G(X_{k-n} - H_{k-n}) \tag{3.147}$$

Consider the Lagrangean expression

$$\mathcal{L} = \sum_{k=0}^{\infty} \{\alpha^k \pi(X_k, H_k) - \lambda_k [X_{k+1} - \sigma(X_k - H_k) - G(X_{k-n} - H_{k-n})]\} \tag{3.148}$$

The first-order necessary conditions for a maximum are

$$\frac{\partial \mathscr{L}}{\partial X_k} = 0 \quad (k \geq 1), \qquad \frac{\partial \mathscr{L}}{\partial H_k} = 0 \quad (k \geq 0) \qquad (3.149)$$

We consider an optimal equilibrium solution with escapement level S:

$$X_k = X = S + G(S), \qquad H = X - S$$

We then obtain, from Eqs. (3.149),

$$\alpha \pi_X + \sigma \lambda_1 + \lambda_{n+1} G'(S) = \lambda_0$$

$$\alpha \pi_H - \sigma \lambda_1 - \lambda_{n+1} G'(S) = 0$$

$$\alpha^2 \pi_X + \sigma \lambda_2 + \lambda_{n+2} G'(S) = \lambda_1$$

$$\alpha^2 \pi_H - \sigma \lambda_2 - \lambda_{n+2} G'(S) = 0$$

and so on (where π_X, π_H denote partial derivatives). Therefore,

$$\lambda_k = \alpha^{k+1}(\pi_X + \pi_H)$$

By substituting into the first equation, we finally obtain

$$[\sigma - \alpha^n G'(S)] \frac{\pi_X + \pi_H}{\pi_H} = \frac{1}{\alpha} \qquad (3.150)$$

and this is the same as Eq. (3.108).

In the case that the delay n is zero and $\pi(X, H)$ has the form

$$\pi(X, H) = \int_{X-H}^{X} f(x) \, dx$$

for some function $f(x)$, the above problem is myopic in the sense discussed in the text. This is easily seen, for if F is an antiderivative of f, we have

$$\sum_{k=0}^{\infty} \alpha^k \pi(X_k, H_k) = \sum_{k=0}^{\infty} \alpha^k [F(X_k) - F(X_k - H_k)]$$

$$= F(X_0) - F(X_0 - H_0)$$

$$+ \sum_{k=1}^{\infty} \alpha^k [F(G(X_{k-1} - H_{k-1}) - F(X_k - H_k)]$$

$$= \sum_{j=0}^{\infty} \alpha^j [\alpha F(G(S_j)) - F(S_j)] + F(X_0)$$

by change of summation index.†

---

†This is the discrete-time analog of integration by parts, which was used to solve the myopic continuous-time model—see the Appendix to Chapter 1.

The solution is now apparent: Escapement $S = S^*$ should be chosen to maximize the expression

$$W(S) = \alpha F(G(S)) - F(S)$$

and the seasonal escapement S_k should be adjusted to the level S^* as rapidly as possible. This is precisely the myopic-equilibrium rule. [Complications arise if the function $W(S)$ above is not convex: see Spence and Starrett (1975), Majumdar and Mitra (1983). Further applications of myopic discrete-time rules are given by Sobel (1982).]

4 Models of Fishery Regulation

In this chapter we use our single-species, deterministic framework to study the effect of a variety of methods of fishery regulation, such as taxes, quotas, and licences. Each of these methods can be applied in a number of different ways—for example, quotas may consist of nonallocated total allowable catches (TACs), or quotas may be allocated to vessels, fishermen, or companies; they can be transferable and marketable, or otherwise; they can be applied to catch, or to effort, or both. The different regulatory methods can also be used in a variety of combinations. Our theory allows for the methodical investigation of these different possibilities.

In order to construct models of fishery regulation, we must first adopt some hypotheses concerning the behavior of individual fishermen. Of course, we assume that the fishermen *compete* for catching fish (subject to appropriate constraints), but the question is whether in so doing, they act in a *strategic* fashion, taking into consideration the actions of competing

143

fishermen, or whether they act "myopically," considering only their own short-term costs and benefits. This matter is studied in Section 4.2.

First, however, we digress briefly to consider possible management objectives.

4.1   Management Objectives

Regulatory policies are presumably introduced with some set of *objectives* in mind, although frequently the original objectives may be quite vague and may later be lost sight of. A selection of possible objectives culled from the fisheries literature is listed in Table 4.1.

Certain objectives seem to be based primarily on biological considerations; sometimes these are referred to-as "scientific" management objectives, perhaps with an underlying implication that sordid economic

TABLE 4.1   Some Possible Objectives of Fishery Regulation

| | Main Purpose | | |
| | | Economic | |
Objective	Biological	Efficiency	Equity
1. Conserve fish stocks	×		
2. Maintain healthy ecosystem	×		
3. Maximize catches	×	×	
4. Stabilize stock levels	×		
5. Stabilize catch rates		×	
6. Provide employment			×
7. Increase fishermen's income		×	×
8. Reduce conflicts			×
9. Protect sports fisheries			×
10. Prevent waste of fish	×	×	
11. Improve quality of fish		×	
12. Maintain low consumer price			×
13. Increase cost-effectiveness		×	
14. Reduce overcapacity		×	
15. Foster development of underutilized stocks	×	×	
16. Increase fish exports		×	
17. Provide government revenue			×

objectives are not worthy of attention or are too vague to be quantified. Actually, most biological objectives do have desirable economic implications, and many of them would be implied by any management policy based on sufficiently broad long-term considerations of social–economic welfare.

The objectives that are primarily of an economic nature fall into two classes, depending on whether the emphasis is on economic efficiency or on the equity of distribution of economic benefits. Whether fishery management should properly be concerned primarily with efficiency or with equity is a topic that has been vigorously debated in academic circles (see, e.g., Scott 1977, Bishop et al. 1981). In real life there is no question that distributional questions tend to dominate the decisions of management agencies, which must deal face to face with the people whose livelihoods are affected by regulations. The result, unfortunately, is often that the overall economic performance of the fishery receives little attention, and consequently is poor by any measure of economic efficiency. This has proven to be the case equally in developed and developing countries, and may even be most extreme in countries that depend critically on fish protein supply (Khoo 1980).

In terms of modelling, the question arises whether one can formulate an optimization model that encompasses most or all of the management objectives listed in Table 4.1. Apart from those which are primarily equity-oriented, most of these objectives can in fact be encompassed by our general "profit-maximizing" sole-owner model—provided the notion of *profit* is replaced by that of *social welfare*. For example, social cost would be used in place of private cost, social utility of consumption would replace monopoly revenue, and the social rate of discount would be used to discount future welfare.

Operational definitions of these socially oriented concepts are given in the literature on welfare economics and social cost/benefit analysis, topics that are beyond the scope of the present work (see Mishan 1971; for a general resources setting, see Herfindahl and Kneese 1974). As an example, however, suppose that unemployment is considered a problem in the fishing sector of the economy, in the sense that potential fishermen are unemployed (or underemployed) and have no viable alternative employment opportunities. Then the *social* cost of employing these fishermen is zero, but since even unemployed fishermen will not work without renumeration, the private cost would be positive. Under these circumstances, the socially optimal level of employment in the fishery may exceed the privately optimal level, and this would be reflected in the social optimization analysis (Munro 1976).

The biologically oriented objectives will usually be achieved if an appropriately low social rate of discount is employed in the analysis. Discount rates in the range 2–5% per annum would seem reasonable here; such rates reflect normal real rates of return on government bonds—although governmental monetary policies in the 1980s have led to fluctuations in real interest rates well beyond this range. Biological objectives which are not readily encompassed by the model can often be expressed instead in terms of additional constraints—for example, the constraint $X \geq X^0$ would imply that, regardless of the outcome of "optimality" calculations, a stock level below X^0 must not be permitted. In practice, quite complicated constraints on exploitation are often imposed as part of management policy.

We maintain, therefore, that most of the nonequity objectives listed in Table 4.1 can routinely be incorporated into our basic welfare-optimization model.

The relationship between efficiency and the distribution of economic benefits is usually modelled in the setting of the theory of cooperative games (Nash 1953). Most of the work in this area, however, has been restricted to static models, for the simple reason that dynamic (e.g., "differential") games are exceedingly difficult to analyze. Except for a brief discussion of a competitive-game fishery model in Section 4.2, we shall not employ game-theoretic methods in this work. Our comparative analysis of the various forms of fishery regulation, however, will include discussion of distributional implications. [See Munro (1979) for a dynamic game-theoretic model of international conflicts.]

4.2   Models of Fishermen's Behavior

For the purposes of the present discussion, the word *fisherman* will be used to refer to the independent owner–operator of a fishing vessel. This is in fact not an unusual situation, although separation of ownership and operation is also common—as are various more complex contractual relationships between vessel owners (or part-owners) and fishermen–operators.

The individual fisherman is faced with a large number of decisions, varying from major capital investments to day-to-day actual fishing operations. Which decisions are actually taken will depend on many factors, including biological factors (relative availability of fish by species and area), economic factors (costs, prices), and institutional factors (regulations and enforcement, tax laws). Assuming that all these factors are known, can we hope to predict what decisions the fisherman will take? Even worse, what can be predicted in circumstances where all the factors are surrounded by uncertainty?

In order to proceed, clearly we must adopt some simplifying hypotheses. The main hypothesis, to begin with, will be that the individual fisherman attempts to maximize his net revenue flow at all times, based on the *current* conditions pertaining to stock abundance, prices, costs, regulations, and so on. Obviously this assumption, like any modelling assumption, must be taken with a grain of salt. One could not hope, for example, that the hypothesis would stand up to empirical testing in any exact sense; fishermen's actual motives will be much more complex than this.

It may not be obvious, but our hypothesis in essence amounts to assuming that the fishery is exploited under *purely competitive* conditions (subject to prevalent regulations). Consider, for example, the decision as to how intensively to fish on a given day. Let E denote the amount of effort exerted for the day by the fisherman; E can be varied in several ways—by fishing longer or shorter hours, at higher or lower running speed, using larger or smaller nets, and so forth. The vessel will have some maximum daily effort capacity E_{max} and the (variable) cost of effort must be some nonlinear function $c(E)$. It is reasonable to assume that marginal cost is positive and increasing, approaching $+\infty$ as $E \rightarrow E_{max}$—see Figure 4.1.

The net revenues from the day's fishing are then

$$\pi = pH - c(E) = pqEX - c(E) \tag{4.1}$$

where it can be assumed that the biomass X does not change perceptibly during one day's fishing. According to our hypothesis, the fisherman

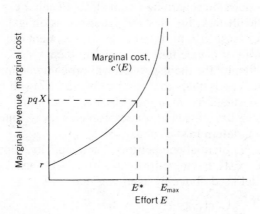

FIGURE 4.1 Marginal cost of daily vessel effort E.

maximizes π, using a daily level of effort E^* given by

$$c'(E^*) = pqX \qquad \text{if } pqX > r$$
$$E^* = 0 \qquad \text{if } pqX < r \qquad (4.2)$$

where $r = c'(0)$. This just says that marginal effort costs equal the marginal revenue, provided $\pi > 0$ (see Fig. 4.1).

Why is this purely competitive? Note that it would never make sense to fish more intensively than E^*, since this would decrease the day's net revenue with no possible later benefits. But the fisherman might consider using $E < E^*$, on the basis that the stock may become depleted if E^* is used continually. Our hypothesis assumes that the fisherman does not consider any such conservation questions, not because he is necessarily shortsighted but because he cannot rely on other fishermen to practice similar resource conservation. In other words, it's a competitive fishery in the usual sense. This myopic competitive fishery hypothesis will be used extensively in the next section, where the effects of regulation will be studied.

Investment Decisions

The hypothesis of myopic behavior seems reasonable as a basis for modelling the day-to-day decisions of competing fishermen, but what about the longer term? For example, under what conditions would the initial decision to purchase a vessel have been made? The capital thus invested is surely largely nonmalleable—otherwise there's no theoretical problem, since all costs can be counted as variable costs. But modelling the fisherman's decision regarding nonmalleable capital expenditures is a much more difficult task, for now the fisherman is obligated to consider the future. He must try to foresee the investment decisions of his competitors—who of course face the same problem.

The main difficulty for the fisherman will probably be in guessing how many competitors he is likely to face. (Of course there will be many other uncertainties, particularly in a new fishery where the total stock is unknown. But we have assumed these problems away, since we are still working within a deterministic framework.)

Let there be N potential competitors. For simplicity, suppose that only one vessel type exists, having effort capacity K^0. Let us also simplify the effort–cost model used above by assuming that

$$c(E) = c \cdot E, \qquad 0 < E < K^0 \qquad (4.3)$$

that is, effort cost is linear up to the vessel capacity K^0. [This can be interpreted as a nonlinear cost function having marginal cost $c'(E) = c$ for $0 \le E \le K^0$ and $c'(K^0) = +\infty$.] Having purchased a vessel and entered the competitive fishery, the fisherman will then use full effort or none:

$$E_t = \begin{cases} K^0 & \text{if } pqX > c \\ 0 & \text{if } pqX < c \end{cases} \tag{4.4}$$

Thus all fishermen will fish at full capacity when $X_t > \bar{X} = c/pq$, and all will stop fishing if $X_t < \bar{X}$. [What happens *at* $X_t = \bar{X}$ is a bit ambiguous in the continuous-time model: total effort must equal the equilibrium level $G(\bar{X})/q\bar{X}$, but how this effort is divided up among the fishermen is not clear. The problem disappears if a discrete-time seasonal model is used; see Section 4.3.]

The discounted net present value of revenues accruing to the individual fisherman is given by

$$\text{NPV} = \int_0^T e^{-\delta t}(pqX_t - c)K^0 \, dt - c_f K^0 \tag{4.5}$$

where $c_f K^0$ is the initial cost of the vessel and T denotes the time at which bionomic equilibrium is reached:

$$\frac{dX_t}{dt} = G(X_t) - qNK^0 X_t, \qquad 0 \le t \le T \tag{4.6}$$

$$X_T = \bar{X} = \frac{c}{pq} \tag{4.7}$$

We can assume that $X_0 > \bar{X}$, for otherwise no investment occurs. The fishery continues for $t > T$, but net revenues are all dissipated.

The biomass trajectory X_t, and hence the time T, are both clearly decreasing functions of N: More vessels deplete the stock more rapidly. (In fact, for sufficiently small N we have $T = +\infty$.) Hence the expression $A(N)$ defined by

$$A(N) = \int_0^T e^{-\delta t}(pqX_t - c) \, dt \tag{4.8}$$

is also a decreasing function of N (see Fig. 4.2); it represents the discounted net revenue per unit of vessel capital, given that N vessels enter the fishery at time $t = 0$. The net present value to each vessel, from Eq. (4.5), is

$$\text{NPV} = [A(N) - c_f]K^0 \tag{4.9}$$

The potential fisherman has to decide, at $t = 0$, whether or not to buy

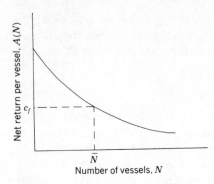

FIGURE 4.2   Net returns per vessel, $A(N)$, in the competitive fishery.

a vessel. If he can make a profit he buys the vessel, otherwise not:

$$\left.\begin{array}{ll} \text{if } A(N) > c_f, & \text{buy vessel} \\[6pt] \text{if } A(N) < c_f, & \text{don't buy vessel} \end{array}\right\} \qquad (4.10)$$

[Note that our formulation automatically takes care of questions such as "Will there be an adequate rate of return on investment?"—just put δ = adequate rate of return and calculate $A(N) - c_f$. The *actual* ("internal") rate of return is that value of δ for which $A(N) = c_f$.]

The fisherman's decision depends critically upon N—but how is he to know what N will be? In an open-access fishery he can only guess. Perhaps he simply sees how many fishermen have already ordered vessels and uses this as his N. If all the fishermen behave this way, and if the decisions are made in rapid sequence, an *investment equilibrium* $\bar{N}$ will occur when

$$A(\bar{N}) = c_f \qquad (4.11)$$

If, on the other hand, investment decisions are spread out over time, $\bar{N}$ will shift (decrease) in response to the progressive depletion of the biomass X_t.

This analysis can be applied to our example of the Antarctic baleen whale fishery, using the parameter values given in Table 3.3. (The equations above are easily modified to allow for depreciation.) The results, corresponding to three values for the initial stock level X_0, are shown in Table 4.2. For example, when $X_0 = 400,000$ BWU (the unexploited biomass), the open-access level of entry $\bar{N}$ given by Eq. (4.8) is 82 factory units (FU). This figure is clearly unrealistic; for example, a fleet of 82 FU would result in fishing the stock down to bionomic equilibrium $\bar{X} = 54,900$ BWU within *one* year (recall that the whale model is linear and assumes infinite price elasticity). Nevertheless, the

TABLE 4.2 Antarctic Whale Model: Optimal and Open-Access Initial Capacity Levels

Initial Biomass X_0 (BWU)	Optimal Initial Capacity K (FU)	Open-Access Initial Capacity $\bar{N}$ (FU)	Time to Reach Equilibrium $\bar{X}$ (years)
400,000	13.5	82	1.0
260,000	9.2	40	1.7
100,000	0	5	∞

extreme profitability of the whale fishery and the corresponding incentive for overcapacity are again emphasized. Even for $X_0 = 260{,}000\,\text{BWU}$ (the "IWC equilibrium"), some 40 FU could enter profitably, in comparison to the 0.66 FU required to harvest the sustainable yield. How does the IWC propose to control whaling when the present moratorium is over and whaling has again become tremendously profitable? As yet, no attention seems to have been paid to this problem.

Restricted Access

It was argued above that competing fishermen in an open-access fishery would act myopically in determining their fishing policy, ignoring the effects of their own catches on the future fish stock. This seems reasonable if access to the fishery is not restricted, but what happens if only a limited number of fishermen are licensed to catch fish? The question is; Would the fishermen, acting independently, be motivated to limit their catches so as to prevent overfishing, or would some form of coercion (mutual or external) still be required? This question is an important one, especially in view of the current trend toward "limited entry" [a phrase that is often used but seldom unambiguously defined—see Rettig and Ginter (1978)].

In fact, there is very little reason to expect that merely restricting the number of licensed fishermen would by itself achieve either biological conservation or economic rationalization of the fishery. There are even theoretical reasons to suggest that the level of overcapacity might *increase* following the introduction of restricted access, and the empirical evidence seems to support such a prediction (Rettig and Ginter 1978).

Let us consider a highly stylized model of the competitive fishing "game" [more complex models are discussed by Clark (1980b), and

Levhari and Mirman (1980)]. Imagine first that just two identical agents ("firms") are granted access to the fishery. Assume that the two agents do not enter into any cooperative agreement on harvest policy and that the government also imposes no other restrictions. Finally, to keep the model as simple as possible, ignore costs of fishing (or assume a type IV concentration profile and ignore fixed costs).

The maximum total profit for the two firms is achieved by maintaining the stock at X^*, where $G'(X^*) = \delta$ (the firms have identical financial opportunities). Suppose $X_0 = X^*$ and consider two possible strategies, "conserve" and "deplete." In the conserve strategy the firms split the optimal sustained yield $H^* = G(X^*)$ 50–50. A firm that chooses the deplete strategy simply takes all the fish it can find.

Table 4.3 is the "payoff matrix" for this simple game; each entry (a, b) in the matrix represents the present values to firm 1 and firm 2, respectively. If both firms employ the conserve strategy, their present-value "profits" are each $pG(X^*)/2\delta$; we set $p = 1$ without affecting the result. If the first firm uses the deplete strategy while the second firm continues to use the conserve strategy, then firm 1 will harvest the entire stock X^* and there will then be nothing left. The table assumes that this depletion is so rapid that the second firm is completely shut out. Finally, if both firms deplete the stock as rapidly as possible, they each catch half of the initial stock X^*.

These are not the only conceivable strategies, but if they were, which would the firms choose? First, it is possible that $X^* > G(X^*)/2\delta$, although for certain $X^* < G(X^*)/\delta$ (given that X^* is really the sole-owner optimum). If so, the depletion strategy does *better* than the conserve strategy, regardless of the strategy that the other firm uses. Hence both firms will be motivated to deplete, and each will actually receive the payoff $X^*/2$, which is *less* than the payoff from mutual conservation.

TABLE 4.3   Payoff Matrix for the Simple Fishing "Game"

	Firm 2	
Firm 1	Conserve	Deplete
Conserve	$\left(\dfrac{G(X^*)}{2\delta}, \dfrac{G(X^*)}{2\delta}\right)$	$(0, X^*)$
Deplete	$(X^*, 0)$	$\left(\dfrac{X^*}{2}, \dfrac{X^*}{2}\right)$

This case is a pure example of the "prisoners' dilemma" discussed in game theory (Luce and Raiffa 1957). The solution (deplete, deplete) is referred to as the *competitive equilibrium* of the game, and it is definitely inferior to the *cooperative* solution (conserve, conserve). Yet the logic of competition seems to force the inferior solution. [Of course, in the case that depletion is "optimal," with $G'(0) = r < \delta$, the two solutions are identical, but this is not the point here.]

In the case that $X^* < G(X^*)/2\delta$, the firm loses revenue by depleting if the other firm tries to conserve, *but the other firm loses more*. If the two firms cannot trust each other—for example, if one thinks that the other is "cheating"—they may be forced to deplete for fear of winding up with nothing. This might be called the "arms race dilemma," since it has the same structure as that notorious phenomenon. It is not certain that the arms race will end in war, nor is it certain that the "fish race" will end in the extinction of the fish, but both are possible outcomes in spite of being obviously completely "irrational."

Now note what happens (in the fishing game) if the number N of competitors is large: The conserve payoff is $G(X^*)/N$ per firm, whereas the "sole cheater" gets X^*, which is obviously larger than his conservation share, provided N is large enough. As N increases, the fishing game passes into a pure prisoners' dilemma, and depletion becomes more likely. Each individual fisherman gains by exceeding his "quota" $G(X^*)/N$ when the future losses are spread over N fishermen. This is the overfishing "externality" argument (or "tragedy of the commons") in its original form.

A simple modification of this game-theoretic model suffices to include variable costs of fishing—in this case "depletion" means that the stock is reduced to $\bar{X} = c/pq$ rather than to zero, but otherwise the arguments are unaffected. Fixed costs, however, are slightly more problematic. The payoff for the deplete strategy depends on the *speed* of depletion, which depends in turn on the firm's fishing capacity.

But this problem has already been discussed: Eq. (4.8) gives the discounted net revenue $A(N)$ per unit of vessel capital given that N units enter the fishery at $t = 0$. If the licensed fishermen (say M in number), are not restricted as to the size of their vessels, and if they adopt the depletion strategy, total capacity $\bar{N}$ will be given by the condition (4.11). The ultimate biomass equilibrium will be either $\bar{X} = c/pq$ (which is reached in a finite time T) or, if fixed costs c_f are large, some higher but suboptimal level (approached asymptotically as $t \to \infty$).

What effect does restricted entry have on this earlier result? This depends on what is meant by "restricted." If, as is assumed here, there are no restrictions other than the exclusion of nonlicensed fishermen (or "firms"), the previous calculations are unaffected. Depletion and overcapacity are virtually certain in spite of the access restriction.

Case studies of restricted entry (see Rettig and Ginter 1978, Sturgess and Meany 1982) have usually indicated that the program has induced an *increase* of fishing capacity—just the opposite of what was presumably intended. In the British Columbia salmon fishery, for example, the number of licenced fishermen was reduced, between 1968 and 1974, by means of a buyback scheme. Yet total capacity increased dramatically during this period, as gillnetters were replaced by purse seiners, and small seiners were replaced by larger, more powerful vessels (Fraser 1980). Similar distortions have been reported in various Australian fisheries (Meany 1978, Hancock 1980) and elsewhere.

This seemingly paradoxical behavior may be explained on the basis of the "option value" of a unit of capacity under restricted entry. Future increases in fish prices will yield benefits to licence holders, in proportion to their fishing capacity. In the absence of access restriction, such price rises would attract additional fishermen and would thus be less profitable to the current fishermen. Restricted access, on the other hand, encourages the fishermen to increase their capacity so as to be in a position to take advantage of such future price increases. The same argument applies to increased total allowable catches, which would be expected if, for example, a stock rehabilitation program were initiated.

To be successful, restricted-entry (licensing) programs must therefore by accompanied by other controls on vessel size, gear, catch quotas, and so on. Conceivably such controls could be introduced (and enforced) by the fishermen themselves through cooperative action, although this seldom seems to have actually happened. The theoretical investigation of these methods is taken up in the next section.

4.3 Regulatory Instruments

The variety of regulations applied to fisheries is impressive. Regulators, in their zeal, have found ways to regulate virtually every aspect of the fishing operation. Regulations affecting vessel type, dimensions, horsepower, and tonnage are common, as are regulations affecting the type, size, and construction of fishing gear. Fishermen may also face regulations pertaining to the time and place of fishing, as well as to species, size, and numbers of fish caught. Processing plants may also be subject to government regulation, mostly of the kind applied generally to food processors.

Because of the central importance of the open-access, common-property problem in fisheries, it will be convenient to adopt a classification of regulatory instruments in terms of the way they affect *exclusiveness* of fishing rights. The arguments of Section 4.2 suggest that this will be a primary consideration in assessing the economic effectiveness of regulations.

Most of the traditional methods of fishery regulation were nonexclusive. The main examples are:

1. *Vessel and gear restrictions*: Physical characteristics of vessels (dimensions, tonnage, horsepower, ancillary equipment) or of fishing gear (type, size, and number of nets, hooks and lines, traps, etc.).

2. *Time and place restrictions*: Fishing seasons, times, and areas.

3. *Catch restrictions*: Species, size, sex of fish caught and retained; restrictions on by-catches, incidental kills, discards.

4. *TACs*: Total quotas by species and area; the fishery is closed when the quota has been achieved.

5. *Quality controls*: Use of ice, freezers, and preservatives; handling and gutting of fish; and so on.

This list is not exhaustive: Further examples include trip limits, citizenship of fishermen, and vessel ownership limitations.

There is no doubt that these methods can be effective in achieving conservation of fish stocks and in increasing the size, quality, and value of catches. The question is: If such regulations are for the good of fishermen, why do they have to be imposed and enforced? Wouldn't the fishermen who adopted them by himself do better than otherwise? This argument still crops up at almost every discussion of fishery management. The reader who is unable to answer it should read this book from page 1 again.

Now what can we predict will happen when all the appropriate regulations have been put in place? Catches will be at a high level and of a high quality. Fishermen will be making good money. By investing some of his income in vessel improvements, the individual fisherman could increase his share of the catch—maybe he will become recognized as a "highliner." Unless the regulations are completely rigid in outlawing such improvements in individual vessels, overall fishing power will increase. If *all* fishermen undertake such "improvements," the net result will simply cancel out—since total catch is already optimized. At the same time, entry to the fishery will tend to expand as new fishermen are attracted by the high profits. Unless this is also brought under control, the fishery will converge to a new "regulated bionomic equilibrium" in which net economic yield is again near zero. Because of the increasing. difficulty and expense of enforcing all the regulations (which will probably appear more and more annoying to the fishermen), the net economic productivity of the fishery may well be lower than if no regulations existed at all. But if regulations are relaxed now, the situation could degenerate into complete disaster.

The next step to be considered, when "too many fishermen are chasing too few fish," is:

6. *License limitation*: Fishing licenses are awarded to a restricted number of fishermen.

This is obviously the first step toward exclusive rights in fisheries. The fishery is thereby transformed from open-access to common-property (Ciriacy–Wantrup and Bishop 1975). Of course, before such a step can be taken, it must be the case either that jurisdiction over the fishery belongs to a single state or that jurisdictional matters have been clearly delineated by international treaty. The establishment of 200-mile Exclusive Fishery Zones in 1976–1977 has been followed by license limitation programs in many areas.

As noted in the preceding section, however, simply restricting access to a finite number of fishermen does not necessarily constitute a solution to the problem of fishery management. Most if not all of the earlier regulations (types 1–5) will still have to be enforced in order to prevent overfishing. Some economists have argued that, provided the regulations are sufficiently comprehensive and enforceable and the number of licenses sufficiently limited, the incentives leading to economic inefficiency can be defeated without further regulations. Others have claimed that fishermen will always be able to outsmart the regulators; the various arguments are well covered in the conference volume edited by Pearse (1979).

Two additional instruments that may be considered are:

7. *Financial disincentives*: Taxes or royalties on catch or on fishing effort or its components.

8. *Quota allocations (quantitative rights)*: Catch quotas are allocated to individual fishermen or to other fishing enterprises.

The bioeconomic implications of these two instruments, which are related in a complementary sense, will be taken up in the next section. At the present time, financial disincentives do not appear to have been adopted except in a few cases involving foreign fleets. Individual fishermen's quotas, however, have been adopted in several fisheries, and in Canada they are under active consideration as a central component of management for marine fisheries on both coasts. The many difficult questions which may have to be addressed before a successful quota allocation scheme can be devised will be discussed in the following section.

4.4 Taxes and Quotas

It is a well-known principle in the economics of regulation that price controls (e.g., taxes) and quantity controls (e.g., quotas) have equivalent effects on production. In a mathematical sense they are "dual" controls, price and quantity being "dual" variables. This principle, however, is by no means universally valid; under conditions of uncertainty, for example, taxes and quotas are not equivalent (see Chapter 6).

Also, in order for the equivalence principle to be valid, the quotas must be allocated to producers and they must be *transferable*. If so, the exchange of quotas will generate a quota market, and the resulting quota price turns out to be equivalent to a tax on production.

Let us see how this works out in our fishery model (Moloney and Pearse 1979, Clark 1980a). Suppose the ith fisherman has a quota Q_i, which allows him to harvest at the rate Q_i:

$$H_i \leq Q_i \tag{4.12}$$

(*Seasonal* quotas will be discussed later.) The total quota Q is fixed by the management authority (but may vary over time):

$$\sum_{i=1}^{N} Q_i = Q \tag{4.13}$$

The quotas are transferable in any portion between the N fishermen (this assumption is crucial).

The fisherman's net revenue is, as before,

$$\pi_i = pqXE_i - c_i(E_i) \tag{4.14}$$

and we have

$$H_i = qXE_i \tag{4.15}$$

where in general $q = q(X)$. Equation (4.14) can also be written in the form

$$\pi_i = \pi_i(X, H_i) = pH_i - \tilde{c}_i(X, H_i) \tag{4.16}$$

where

$$\tilde{c}_i(X, H_i) = c_i\left(\frac{H_i}{qX}\right) \tag{4.17}$$

We assume that the fishery is competitive and that the fishermen are not able to surpass their quotas. Hence each fisherman attempts to maximize π_i subject to his quota:

$$\max_{H_i \leq Q_i} \pi_i(X, H_i) \tag{4.18}$$

The fisherman can also buy or sell quota units. Let m denote the price on the quota market. Clearly the fisherman will not retain unused quota units (this is where the deterministic assumption is most important!); he could profit from purchasing an additional unit if and only if

$$\frac{\partial \pi_i}{\partial H_i}(X, Q_i) > m$$

Consequently, the equation

$$\frac{\partial \pi_i}{\partial H_i}(X, D_i) = m \tag{4.19}$$

determines the ith fisherman's *demand function* $D_i = D_i(m; X)$ for quota units. Explicitly, this becomes

$$c_i'\left(\frac{D_i}{qX}\right) = (p - m)qX \tag{4.20}$$

and from this it follows readily that

$$\frac{\partial D_i}{\partial m} < 0 \quad \text{and} \quad \frac{\partial D_i}{\partial X} \geq 0 \tag{4.21}$$

(and in fact $\partial D_i/\partial X > 0$ except for type IV concentration). These inequalities agree with intuition.

The total demand for quotas is

$$D(X, m) = \sum_{i=1}^{N} D_i(X, m) \tag{4.22}$$

The supply–demand equilibrium condition

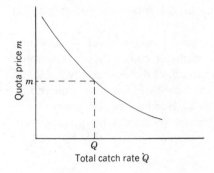

FIGURE 4.3 Demand for quotas and the price of quotas.

$$D(X, m) = Q \qquad (4.23)$$

thus determines the market-clearing quota price m—see Figure 4.3. We have $dm/dX = -(\partial D/\partial X) \div (\partial D/\partial m) \geq 0$ (>0 except for type IV): Higher stock levels X imply that quotas are more valuable because harvest costs are lower.

When quota transfers are completed (the quota market has "cleared") we have $H_i = Q_i = D_i$, and Eq. (4.20) becomes

$$c_i'(E_i) = (p - m)qX \qquad (4.24)$$

It is this equation that demonstrates the equivalence of catch quotas and catch taxes. For suppose the fisherman must pay a tax τ on his catch. His after-tax revenue is then

$$\pi_{i\tau} = (p - \tau)H_i - c_i(E_i) \qquad (4.25)$$

maximization of which implies that

$$c_i'(E_i) = (p - \tau)qX \qquad (4.26)$$

Thus E_i is determined as a function of X and τ; clearly,

$$\frac{\partial E_i}{\partial \tau} < 0 \qquad (4.27)$$

Hence the total effort $\sum_{i=1}^{N} E_i$ is also a decreasing function of the tax rate τ; for $\tau \geq p$, obviously $E_i = 0$.

By comparison of Eqs. (4.24) and (4.26) we see that a tax τ has the same effect on fishing effort as a transferable quota with price m. This equivalence can also be understood as follows: The tax is a direct cost to the fisherman and actually reduces the price received for his catch; on the other hand, the transferable quota represents an *opportunity* cost, since rather than using his marginal quota, the fisherman could sell it at the price m.

Equation (4.27) implies that, given any biomass level X and any desired total effort level E, the fishery manager can in theory determine a tax τ on catch which will cause effort to equal the desired level. Hence any desired time schedule of effort E_t (and catch H_t) can be achieved by means of an appropriate tax schedule τ_t. Because of the equivalence principle, the same result can be achieved by means of an allocated, transferable quota scheme.

Optimization

According to this theory, either a system of allocated transferable catch quotas or a schedule of taxes on catch can be used by the fishery manager to achieve any desired schedule of total catch rates. (The quotas or taxes could also be imposed on effort with the same result, since our present model assumes a one-to-one relationship between catch rate and effort. This assumption again breaks down in a stochastic world.)

A *nonallocated* total catch quota ("TAC") system could also be used to control catch rates, of course. The implications for economic efficiency, however, are completely different for allocated and nonallocated quota systems. This is intuitively clear, since a nonallocated quota system forces fishermen to compete furiously for a share of the catch, while the fisherman who owns a personal quota can fish at a more leisurely—and more efficient—pace. In order to analyze this situation, we consider a more realistic seasonal model of the fishery (see Section 3.6).

For simplicity, assume that all fishermen have identical cost-of-effort functions $c(E)$ [the case of different cost functions $c_i(E_i)$ is easily dealt with, but slightly messier—see Clark (1980a)]. The intraseasonal optimization model is

$$\text{Maximize}_{N, E_t} \left\{ N \int_0^T [pqxE_t - c(E_t)]\, dt - c_f N \right\} \qquad (4.28)$$

subject to

$$\frac{dx_t}{dt} = -qx_t NE_t, \qquad 0 \le t \le T \qquad (4.29)$$

$$x_0 = R, \qquad x_T = S \qquad (4.30)$$

where R and S, which respectively denote recruitment and escapement for the given season, are first treated as parameters. In Eq. (4.28), c_f denotes seasonal fixed cost ("mobilization" or "setup" cost). Long-term fixed costs of nonmalleable vessel capital are ignored for the moment.

The above optimal control problem can be solved by the maximum principle (see the Appendix to Chapter 3); the Hamiltonian expression is

$$\mathcal{H} = N[pqxE - c(E)] - c_f N - \lambda qxNE \qquad (4.31)$$

and the necessary conditions for optimality are

$$\frac{\partial \mathcal{H}}{\partial E} = 0 \qquad (4.32)$$

$$\frac{d\lambda}{dt} = -\frac{\partial \mathcal{H}}{\partial x} \qquad (4.33)$$

Also $\mathcal{H}$ must be maximized with respect to N. Writing out the calculations, we obtain

$$c'(E_t) = (p - \lambda_t)qx_t \qquad (4.34)$$

$$\frac{d\lambda_t}{dt} = N(p - \lambda_t)E_t \frac{d}{dx}(qx) \qquad (4.35)$$

From Eqs. (4.35) and (4.29) we obtain

$$\frac{d}{dt}[(p - \lambda)x] = (p - \lambda)\frac{dx}{dt} - x\frac{d\lambda}{dt}$$

$$= (p - \lambda)NEx^2 \frac{dq(x)}{dx}$$

In the case that catchability is constant (type II), the right side vanishes, and we conclude that

$$(p - \lambda)qx = \text{constant}$$

Then Eq. (4.34) implies that $E = \text{constant}$ also—each vessel should employ a constant level of effort throughout the fishing season. A type I concentration profile, on the other hand, would require optimal effort to increase as the fishing season progresses; the reverse would apply to types III and IV.

It seems unlikely, for most fisheries, that catchability would vary significantly *within* a given fishing season, although variation over several seasons could well occur. Let us therefore suppose that $q = \text{constant}$ for the given season. Then $E = \text{constant}$, and we have from Eq. (4.29)

$$NE = \frac{1}{qT} \ln \frac{R}{S} \qquad (4.36)$$

where NE is the total seasonal effort.

Now E must be chosen to maximize $\mathcal{H}$ in Eq. (4.31). Since NE is fixed

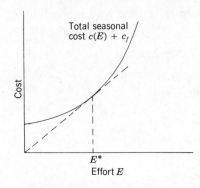

FIGURE 4.4 Optimal vessel effort E^* for the seasonal model.

by (4.36), this implies that E must minimize the terms

$$Nc(E) + c_f N = a\,\frac{c(E) + c_f}{E} \tag{4.37}$$

where the constant a denotes the right side of (4.36). But this says simply that the optimal effort per vessel, E^*, should minimize average cost, including seasonal fixed cost; a diagrammatic solution is shown in Figure 4.4 (see also Anderson (1977). The optimal number of vessels, N^*, as given by Eq. (4.36), is just sufficient to catch the seasonal quota $R - S$.

The details of the optimal solution are not particularly relevant at this stage; we are mainly interested in Eq. (4.34):

$$c'(E^*) = (p - \lambda_t)qx_t$$

Thus, by imposing a tax $\tau_t = \lambda_t$ on catch, the management authority can force the *competitive* fishery into the optimal mode; *total* effort NE^* will be optimal provided the number of vessel licenses N is optimal. Since the biomass x_t is given by

$$x_t = \mathrm{Re}^{-qN^*E^*t}$$

the optimizing tax can be determined explicitly:

$$\tau_t = p - \frac{c'(E^*)}{qx_t} = p - \frac{c'(E^*)}{qR}\,e^{qN^*E^*t} \tag{4.38}$$

The tax thus *decreases* as the fishing season progresses. The fishermen's net revenue flow, however, remains constant:

$$\pi_{\text{net}} = (p - \tau_t)qx_t E^* - c(E^*)$$
$$= c'(E^*)E^* - c(E^*)$$

This mathematically elegant solution is doubtlessly a bit too elegant for practical use—varying the tax τ over the season would be difficult, and the fishermen would tend to store their catch until lower tax rates prevailed, thus thwarting the scheme. In most cases, the compromise of a constant tax $\bar{\tau}$ would be fairly close to optimal, the exception being when the stock x_t is significantly changed by fishing each season.

From the equivalence of taxes and allocated quotas, it follows that the optimal quota Q_t would also vary during the season; in fact, we have

$$Q_t = NqE^*x_t$$

so that the quota should also decrease. Here Q_t denotes a quota "rate," not a seasonal quota. In practice, such a quota would be realized by short-term (e.g., weekly) quotas. In most actual cases, however, variations in the quota from week to week would not be worthwhile, and an overall allocated seasonal quota would suffice. (Weekly allocated vessel quotas have been employed, since 1979, in the Bay of Fundy purse-seine herring fishery. The original seasonal allocated quota was divided into weekly units in order to smooth out the supply of fish to processing plants.)

Next we consider the case in which the individual fishermen's quotas are not transferable. Suppose the fisherman has a fixed seasonal quota $Q^0 \ll Q =$ total quota for the N fishermen. The fisherman's own catches are a negligible portion of the total catch, and he therefore considers the fish stock x_t as an exogenous variable. He attempts to maximize his seasonal net revenue:

$$\underset{E_t \geq 0}{\text{Maximize}} \int_0^T [pqx_tE_t - c(E_t)]\, dt \qquad (4.39)$$

subject to the quota constraint

$$\int_0^T qx_tE_t\, dt = Q^0 \qquad (4.40)$$

(we assume that the quota actually influences the fisherman!). Equation (4.40) can be written in the form

$$\frac{dy_t}{dt} = qx_tE_t, \qquad y_0 = 0, \qquad y_T = Q^0 \qquad (4.41)$$

where y_t represents the cumulative catch up to time t. In this form, the maximization problem is a standard optimal control problem, and is easily solved by the maximum principle. We immediately obtain

$$c'(E_t) = (p - \mu_t)qx_t \qquad (4.42)$$

$$\frac{d\mu_t}{dt} = 0, \qquad \text{i.e., } \mu_t = \text{constant} \qquad (4.43)$$

This has the same form as the optimality rule, Eq. (4.34), except that the "shadow price" λ_t is replaced by a constant, μ_t. This implies that the fisherman will use a decreasing level of effort E_t—he will tend to fish most intensively at the beginning of the season when the fish are most abundant and catch per unit effort is highest. As noted above, this behavior is not theoretically optimal, but in most cases the difference would perhaps be academic. (If not, the quota could be subdivided into short-term units.) Thus an allocated, nontransferable quota system is capable of generating approximately optimal fishing effort—provided each fisherman is allocated the "correct" quota to begin with. If different fishermen have different cost functions $c_i(E_i)$, then the appropriate quotas will likewise be different. The transferable quota market would doubtlessly be the superior method of achieving the proper quota distribution in this case.

The results so far are summarized in Table 4.4. Two of the entries in the table have not yet been discussed. The case of a tax on effort is easily worked out; more interesting is the possibility of combining taxes and quotas. Assuming that the quotas are allocated and transferable, it is easy to see that the quota price will simply be reduced by the amount of the tax:

$$m'_t = m_t - \tau_t \tag{4.44}$$

TABLE 4.4 Marginal Cost-of-Effort Equations for Various Management Regimes

Management Regime	Marginal Cost Equation	Interpretation
Social optimum (or sole owner)	$c'(E) = (p - \lambda_t)qx_t$	$\lambda_t =$ "shadow price"
Tax on catch	$c'(E) = (p - \tau_t)qx_t$	$\tau_t =$ catch tax
Tax on effort	$c'(E) = pqx_t - \epsilon_t$	$\epsilon_t =$ effort tax
Allocated, transferable catch quota	$c'(E) = (p - m_t)qx_t$	$m_t =$ quota price
Allocated, transferable catch quota plus tax on catch	$c'(E) = (p - \tau_t - m'_t)qx_t$	$\tau_t =$ catch tax, $m'_t =$ quota price
Allocated, nontransferable catch quota	$c'(E) = (p - \mu)qx_t$	$\mu =$ constant
Nonallocated total catch quota	$c'(E) = pqx_t$	—
Unregulated	$c'(E) = pqx_t$	—

Distributional Implications

While taxes and allocated quotas may be equivalent at the theoretical level assumed in the previous discussion, in practice they obviously differ in important respects. In particular, their distributional effects are quite dissimilar. Taxes of the magnitude necessary to prevent overfishing must remove all of the economic rent from the fishery. If the fishery is already depleted, the taxes must be large enough to drive some existing fishermen out of the fishery entirely. The unpopularity of such measures requires no comment.

Allocated quotas, on the other hand, leave the economic rent in the hands of the quota holders. If transferable, the quotas will acquire a value reflecting this rent. Indeed, nontransferability of fishing rights has proven to be difficult to administer, precisely because of the value inherent in such rights.

The fact that allocated quotas are valuable means that the allocation process itself is largely a matter of awarding favors. The allocation process is therefore clearly subject to political interference, and even to the influence of corruption. In the Alaskan salmon fishery, for example, the establishment of exclusive fishing privileges has been accompanied by years of court battles and litigation (Adasiak 1978, Bishop et al. 1981).

It is also a question of social justice whether valuable resource rights should be given away gratis to privileged individuals, no matter how deserving. In other resource industries, exploitation rights are often sold to the highest bidder, but political exigencies appear to have inhibited the application of this rule to fisheries. [An exception, albeit very small, occurs in the San Francisco Bay roe herring fishery, where quotas are sold at auction each season; see Huppert (1982).]

The possibility of combining taxes with quotas or other forms of fishing rights provides the government with the opportunity of achieving both efficiency and the desired distribution of economic benefits. As shown by Eq. (4.44), the tax can be used to transfer any desired portion of quota values to the government.

Processors and Consumers

In order to keep the above arguments simple we have again ignored both consumer surplus and the existence of a processing sector. These considerations can be reintroduced into the model, at the cost of a certain amount of complexity and tediousness, but we shall not attempt to go into all the details.

For example, it follows as in Section 3.4 that if the fish-harvesting sector is competitive, then economic optimization could be achieved by means of taxes or quotas, or a combination thereof, applied to the processing sector. In the case that the processing sector is monopsonistic, it is possible in theory that a *subsidy* (i.e., a negative tax) would be required, although the possibility seems remote (see Clark and Munro 1980).

One question that is often raised is whether taxes on fishermen (or processing firms) will be reflected in increased prices to consumers. There are many possibilities, depending on the effect of taxes on catch rates, the elasticity of demand, and the monopoly position of the industry. In practice, however, taxes will usually only be considered in a fully developed fishery in which total catches are already completely controlled. In this case neither the supply nor the demand for fish is affected by the tax, and there should be no effect on consumer prices.

Capitalization and Innovation

The foregoing analysis has been concerned solely with the use of taxes or allocated quotas to improve the intraseasonal input of fishing effort. Either method has the effect of smoothing effort out over the fishing season, thereby reducing the costs of catching the season's quota. The quota itself is either fixed directly, in the case of allocated quotas, or is achieved (in theory) as a consequence of the tax. Other advantages not directly included in the model might include reduction in processing and storage costs and improved quality of the product resulting from better on-board handling and storage of fish.

The use of taxes or allocated quotas to control overexploitation and to reduce fixed costs has yet to be assessed. As shown in Section 3.3, the question of fixed costs is relevant only when investment is irreversible. Rather than attempting to analyze this question mathematically, we present a numerical illustration.

Imagine a regulated fishery with the following characteristics:

Number of licences:	$N = 100$
TAC:	$Q = 100{,}000$ tons/year
Total effort required:	$E = 10{,}000$ SFD/year
Variable cost of effort:	$c = \$2000$/SFD
Annual fixed (setup) cost:	$c_f = \$100{,}000$ per vessel
Price of landed fish:	$p = \$500$/ton
Individual vessel capacity:	$E_i^{max} = 200$ SFD/year

(Here SFD denotes Standardized Fishing Day.) Catch per unit effort remains constant at 10 tons/SFD throughout the season. The vessels are identical and have no alternative uses, so that vessel capital is completely nonmalleable. The total fleet capacity of 20,000 SFD/year is twice the required capacity.

In working with this "tabular" model (a popular form with the early economists) we will assume that individual vessel owners have perfect information and always act so as to maximize their annual income.

Assume first that fishing is competitive. All vessels fish at maximum effort until the TAC is taken. Each vessel employs 100 SFD and catches 1000 tons/year. Annual costs and revenues per vessel are:

Variable cost:	$2000 \times 100 =$	$200,000
Fixed (setup) costs:		100,000
Total cost:		300,000
Revenue:	$500 \times 1000 =$	500,000
Net income:		$200,000

Next consider a system of transferable quotas, with each licence owner allocated a quota of 1000 tons. One additional ton of quota would be worth $300, so that (marginal) quotas would trade at $m = \$300/\text{ton}$. Because of the fixed costs, however, the value of the *entire* quota of 1000 tons is only $200,000. A fisherman would therefore sell his entire licence plus annual 1000-ton quota at a price that would yield him an income of at least $200,000 per/year.

On the other hand, by buying up another vessel's quota, a fisherman could increase his income by $300,000/year. This original situation is obviously unstable; 50 vessels will be retired from the fishery, transferring their quotas to the remaining 50 vessels. The transferable quota system therefore results in a reduction of capacity, leading to an annual saving in setup costs amounting to $5 million. (Who gets the $5 million depends on the prices negotiated between buyers and sellers of the quotas.) Total industry net annual revenue under the two systems is:

	TAC	Allocated Quota
Variable cost	$20 million	$20 million
Fixed setup cost	10 million	5 million
Revenue	50 million	50 million
Net income	$20 million	$25 million

Note that the $5 million saving comes entirely from the reduction in fixed "setup" costs of $100,000 per vessel. Costs of this kind would mainly be associated with sailing from port to the fishing ground and back. Such costs can be substantial.

A government buyback program removing 50 vessels would cost the government at least $50 \times \$200,000 = \$10$ million (per annum equivalent), and this would constitute an outright gift to the remaining 50 fishermen.

Next we wish to consider the introduction of a capital-intensive innovation—for example, a new vessel type or a new kind of gear. The innovation permits the fisherman to double his daily catch at the same cost per day; thus vessel capacity is increased to 400 SFD/year and the variable cost of effort is reduced from $2000 to $1000/SFD. Interest and depreciation on the cost of the innovation amounts to $W/year. We wish to calculate, as a function of W, the number of fishermen who will adopt the innovation, under five sets of circumstances:

1. Optimal
2. Competitive fishery, with no quota allocations
3. Transferable allocated quotas
4. Nontransferable allocated quotas
5. A tax on catch

Other possibilities can be treated by similar arguments.

The optimal policy is either to replace all the old vessels with 25 new vessels or to retain 50 old vessels. Net annual revenue for the industry is $37.5 million and $25 million, respectively. Hence the new vessels should be introduced if $W < (\$37.5 \text{ million}-\$25 \text{ million}) \div 25 = \$500,000$, and not otherwise:

$$N^* = \begin{cases} 25 & \text{if } W < \$500,000 \\ 0 & \text{if } W > \$500,000 \end{cases}$$

In the competitive case, suppose n fishermen purchase new vessels, the remaining $100 - n$ retaining their old vessels. If x denotes the annual catch of an old vessel, then the catch of a new vessel is $2x$, and we have $2nx + (100 - n)x = Q = 100,000$ tons, or

$$x = \frac{100,000}{n + 100} \text{ tons}$$

Similarly, if $y =$ number of days fishing, then $2ny + (100 - n)y = 10,000$, or

$$y = \frac{10,000}{n+100} \text{ days}$$

Net annual income per vessel is then

New vessels: $\$8000y - 100,000$
Old vessels: $\$3000y - 100,000$

If we assume, as in Section 4.2, that the fishermen correctly forecast the number $\bar{n}$ of new vessels, then $\bar{n}$ will be such that the additional income per vessel equals W; thus $5000y = W$, or, substituting for y,

$$\bar{n} = \frac{50 \text{ million}}{W} - 100$$

For $W < 250,000$ all fishermen replace their vessels and total fishing capacity doubles. Otherwise, only a partial replacement occurs:

W	$\bar{n}$
$500,000	0
400,000	25
300,000	66
250,000	100

Next consider the case of allocated transferable quotas. The reduction to 50 old vessels has already taken place. The new vessel owner buys up the quotas/licences of two old vessels to obtain a quote of 4000 tons/year. Net annual income per vessel is

New vessels: $1.5 million
Old vessels: $0.5 million

Thus it costs $1 million to buy up the two licences, and the net increase is $500,000/year. Hence no new vessels will enter if $W > \$500,000$, and 25 new vessels will replace the 50 old vessels if $W < \$500,000$. The allocated, transferable catch quota system therefore serves to optimize the introduction of innovations.

Consider next the case of nontransferable quotas. Under the original assumptions, each of the 100 fisherman has a quota of 1000 tons/year, yielding a net annual income of $200,000. If the fisherman buys a new vessel, he reduces his variable cost by 50%, and annual income increases to $300,000. Thus the new vessels would not be brought in unless $W < \$100,000$, but all 100 vessels would be replaced if $W < \$100,000$.

Nontransferability inhibits worthwhile innovations unless they are inexpensive, but in that case encourages excess innovation.

Alternatively, suppose the number of licences has been reduced to 50, each with a 2000-ton quota (the optimum for old vessels). The net annual incomes for old and new vessels then become \$500,000 and \$700,000, respectively. The critical value of W is now \$200,000, but the conclusions are similar to the previous case.

Finally, consider a tax τ (\$/ton) on catch; there are no quota allocations. For the preinnovation fishery, each fisherman's income is reduced by \$1000τ; if $\tau > 200$, net income becomes negative and some of the fishermen will withdraw from the fishery. If n fishermen remain, net income per vessel is easily seen to be

$$\$100,000 \left(\frac{300 - \tau}{n} - 1 \right) \qquad \text{if } n \geq 50$$

$$\$500,000 - 2000\tau \qquad \text{if } n < 50$$

(since when $n < 50$, the vessel capacity of 200 SFD limits each vessel's catch). Thus the number n of vessels remaining in the fishery is given by

$$n = \begin{cases} 100 & \text{if } \tau \leq 200 \\ 300 - \tau & \text{if } 200 < \tau < 250 \\ 50 & \text{if } \tau = 250 \\ 0 & \text{if } \tau > 250 \end{cases}$$

A tax of \$250/ton results in the optimum number of vessels (the "knife-edge" discontinuity for $\tau > 250$ is only an artifact of the model, caused by the capacity assumption $E^{\max} = 200$ SFD).

Next consider new vessels; assume $\tau = \$250$/ton. Suppose m new vessels enter and n old vessels remain. The catch x and effort y of an old vessel are then

$$x = \frac{100,000}{n + 2m} \text{ tons}, \qquad y = \frac{10,000}{n + 2m} \text{ SFD}$$

Net revenue for an old vessel equals

$$\$100,000 \left(\frac{300 - \tau}{n + 2m} - 1 \right)$$

so that

$$n + 2m = 50$$

Thus each new vessel displaces two old vessels. Net revenue for a new vessel equals

TABLE 4.5   Summary of the Numerical Illustration

Management Regime	Number of Vessels	
	Old	New
Optimum	50	0 or 25 if $W \geqq \$500,000$
Competitive TAC	100	$5 \times 10^7 / W - 100$
Transferable allocated quota	50	0 or 25 if $W \geqq \$500,000$
Nontransferable allocated quota	100	0 or 100 if $W \geqq \$100,000$
Nontransferable allocated quota (50 licences)	50	0 or 50 if $W \geqq \$200,000$
Tax, \$250/ton	50	0 or 25 if $W \geqq \$500,000$

$$\$100,000 \left(\frac{800 - 2\tau}{n + 2m} - 1 \right) = \$500,000$$

Therefore, if $W < \$500,000$, a total of 25 new vessels will enter the fishery, displacing all the old vessels, whereas if $W > \$500,000$, no new vessels will enter. We conclude that the \$250/ton tax also optimizes the introduction of innovations.

The results of the numerical illustration are summarized in Table 4.5.

To conclude, the numerical illustration† suggests that our previous results can be extended to cover capitalization and innovation. Either transferable allocated catch quotas or appropriate catch taxes (or a combination thereof) can be used to achieve the economically optimal operation of the fishery, and either will also encourage an optimal degree of innovation and introduction of new capital. Other forms of regulation, such as TACs or nontransferable quotas, optimize neither operation nor innovation. [An analytic treatment of the problem of capital costs in an open-access fishery appears in McKelvey (1984).]

---

†The reader may wish to decide whether the illustration is an artificial one! Or he or she may wish to extend it—for example, by introducing explicit figures for wages and the opportunity cost of labor.

Information Requirements

Any management regime must obviously be based on sound scientific estimates of stock abundance and production. Additional types of information will be required if sophisticated, economically oriented management objectives and techniques are to be adopted.

In order to compute the correct tax, for example, the management authority must know the details of costs and prices. Changes over time must be kept track of and the tax adjusted accordingly.

Under an allocated transferable quota system, the authority must keep tabs on quotas and catches by each quota holder. Since the total quota will usually be determined primarily by biological considerations, detailed information on costs and prices will not usually be required. Thus information requirements for the quota system will usually be significantly less than for the tax method; in essence, the quota system serves to *decentralize* economic decisions and the corresponding need for information.

Some further aspects of information in fisheries will be discussed in Chapter 6.

Enforcement

A well-managed fish stock constitutes a valuable asset, which must be protected. Significant economic benefits can accrue to anyone who is able to catch fish illegally and avoid detection. Violations of fishery regulations are common, sometimes reaching extreme proportions.

For example, in 1975 Canadian fisheries officers uncovered a Mafia-style organization pilfering migrating salmon from the Fraser River, where catches are illegal (except for a small native food fishery). The fish were caught and transported to illegal markets at night. The operation was so efficient that almost no salmon escaped to the upriver spawning grounds—in fact, it was the failure of spawners to show up that alerted the authorities to the illegal operation.

Enforcement problems in international fisheries are if anything even more difficult. In establishing 200-mile fishing zones in 1976, the Canadian government cited as one of the leading motivations the incessant violation by Soviet trawler fleets of quotas established and agreed upon by the International Council of Northwest Atlantic Fisheries. In Malaysia, common infractions include poaching by domestic and foreign fishermen, use of illegal gear, and bribing of fishery officers. Clashes between inshore and offshore trawler fishermen resulted in 34 deaths in the area between 1964 and 1976 (Khoo 1980). Similar problems are reported throughout the southwest Pacific.

Returning to domestic fisheries, where national jurisdiction should in theory provide a basis for the enforcement of fishery regulations, we must still recognize that enforcement may be difficult and costly. Fishery closures and similar methods are sometimes fairly easily enforced, although violations are widely reported. More complex regulations engender more difficult enforcement problems, but if enforcement is ignored, the entire management system may be put in jeopardy.

Fishermen would have an incentive to thwart either allocated catch quotas or catch taxes by disposing of their catch through illegal channels. Such programs would therefore have to be accompanied by careful control of the landing and marketing of the catch. Government inspection of all landings would probably be necessary.

It is important for the fishermen to know that the management regulations are strictly enforced and violators punished. Under any licensing program, the threat of losing one's licence would provide a strong incentive for adhering to the regulations. This alone constitutes a strong argument in favor of limited-entry and licensing programs.

Many management regimes seem to degenerate into a perpetual conflict between the fishermen and the management authorities. But in the state of Michigan, where the commercial Great Lakes fisheries are now managed under a rigorous program which includes marketable quotas, the fishermen are reported to have switched to become strong supporters of the management system (D. Jester, personal communication; see also Talhelm 1978).

Implementation of Allocated Quota Systems

Although allocated catch quotas are still a relatively new concept in fisheries management, they seem to offer the best practical hope for the economic rationalization of commercial fisheries. Implementing a successful quota system will not be a simple task, however. Some fisheries may be so complex, in terms of stocks, species, geographical area, gear types, markets, and so on, that no allocated quota system will prove feasible. But all fisheries are complex, and no management system is likely to be ideal. As experience with carefully planned and administered quota systems increases, it is to be expected that more fisheries will be brought under this form of management.

It may not be necessary to break down quota allocations to the level of the individual vessel or fisherman. If the fishery is exploited by several licensed fishing companies or other organized groups of fishermen, government allocation of catches to this level may be appropriate.

For example, Canada's Atlantic offshore groundfish fisheries were placed under a system of "enterprise quotas" beginning in 1982 (Canada Department of Fisheries and Oceans 1983). Total allowable catches of nine species of groundfish (cod, haddock, redfish, American plaice, yellowtail, witch, flounder, Greenland halibut, and pollock) were allocated, on a historical basis and for a 5-year period, to the six existing fishing companies or company groups. Designed with the cooperation of the industry and subject to annual review and revision, the management system allows for the permanent transfer or sale of quotas as well as for season-by-season transfers. Certain regulations on vessel replacement were introduced, but it is hoped that these will become unnecessary as each company attempts to optimize its own vessel use. Company officials have expressed optimism that the enterprise quota system will encourage stability and economic efficiency in these important fisheries. The Canadian Fisheries Department has announced its intention of gradually introducing similar systems into the inshore fisheries.

4.5 Summary

The fundamental reason for fishery regulation is the common-property nature of the resource. If the fishing industry is more difficult to regulate than other industries, it is largely because the kinds of property-rights delineations common in other situations do not readily apply to fish in the sea (Pearse 1980, Scott 1982).

Traditional methods of fishery management, such as TACs and fishery closures, bypass the property-rights problem by ignoring it. These methods inevitably result in economic distortions which lead to further increases in fishing capacity and hence tend to render control of the fishery progressively more difficult. They have been categorized as "regulation by maximization of inefficiency."

The first break away from these traditional approaches has been toward "limited-entry" programs, wherein fishing privileges are restricted to a specific group of individuals, usually through the issue of licenses. By itself, such a program does not directly address the common-property problem. Although simpleminded models seem to have suggested that limitation of entry would lead to a desirable reduction in "effort," both theory and experience indicate that this hope is largely unfounded. Each license holder is motivated to increase his fishing capacity to compete for a larger share of the catch; in fact, the "option value" argument indicates that ultimate capacity may be greater under licensing than without it.

In order to overcome these difficulties, some method of attaching rights (or proxies for rights) to the resource itself must be devised. The two available methods (other than sole private ownership) are allocated catch quotas, which vest resource rights in the quota holder, and catch taxes, which place the resource rights in the hands of the state. These may be combined, in which case the rights will be split. At the level of abstraction adopted in this chapter, namely deterministic, single-species models, taxes, and *transferable* quotas are shown to be economically equivalent. But as the discussion of information requirements suggests, quotas and taxes would have quite different implications in the real world—quite apart from the distributional question. We will return to this matter in the following chapters.

5 Models of Multispecies Fisheries

Even though modern fishing vessels are often highly specialized, it is rare to find a vessel that exploits a single species of fish. Tuna purse seiners catch both yellowfin and skipjack tuna; pelagic whale fleets took blue, fin, sei, minke, humpback, and sperm whales; salmon fleets in the northeast Pacific exploit five species of salmon consisting of several thousand genetically and geographically separable races.

In the northwest Atlantic over 30 species of groundfish are exploited commercially. The extreme is doubtlessly reached in tropical fisheries, where the catch of a day's trawling may contain as many as 200 species. Overfishing often shows up in such multispecies trawl fisheries in terms of progressive depletion of the more valuable species, which are often those at the higher trophic levels (Pauly 1982, Pauly and Murphy 1982).

In some cases, incidental catches may lead to depletion of nontarget species. Examples among marine mammals include some seven species of

176

porpoises captured in tuna purse seines, and dugongs (similar species to manatees) entrapped in shark nets (Heinsohn 1972). Long-liners have been accused of killing significant numbers of sports fish such as marlin and sailfish.

On the other hand, species at different trophic levels may be exploited by different vessels and gear, examples being the cod–capelin stocks of the North Atlantic and the krill–whale stocks in both northern and southern oceans. In such cases, large catches from the lower trophic level can have serious implications for production at both levels. Seldom, however, is sufficient information available to formulate quantitative predictions of the effects of given levels of catch (Akenhead et al. 1982).

Important fish stocks that have collapsed under fishing pressure have sometimes been replaced by other species which appear to fill the same ecological niche. In the case of the California current system, the replacement of the Pacific sardine by northern anchovy has been shown to occur naturally with a period of about 600 years, but the latest switch seems to have been precipitated in the 1950s by overfishing of the sardine (Murphy 1977). Similar fishery-induced replacements appear to have occurred in the upwelling systems of the southeastern Pacific and Atlantic Oceans.

Another phenomenon of importance in certain multispecies fisheries is the problem of incidental catches or *by-catches*. Catches of unwanted ("trash") fish by shrimp trawlers, for example, often exceed the catch of shrimp by weight. Most of this by-catch consists of juveniles of large fin-fish species, some of which may be of commercial importance. Groundfish trawlers in the northeast Pacific capture significant quantities of Pacific halibut (which they are not permitted to retain), thereby reducing the catches available to halibut vessels. In most cases the mortality rate for incidental catches is nearly 100%. Nevertheless, a rule prohibiting the retention of individual catches (for example, as in the case of halibut) at least inhibits targetting on that species.

Management of multispecies fisheries is rendered difficult both by the complexity of the situation and by the irreducible uncertainty regarding ecological interrelationships in the ocean. The traditional maximum-sustainable-yield (MSY) approach to fishery management offers no useful guidelines when it comes to specifying objectives for the management of interrelated species.

One of the first attempts to introduce something more sophisticated than the MSY objective into an international treaty occurred in the 1980 Convention for the Conservation of Antarctic Marine Living Resources. Many species in the Antarctic ecosystem, inluding whales and other marine mammals, fish, and birds, feed upon the immense krill population, and ecologists have expressed the concern that the imminent development of a large-scale krill fishery could upset the entire ecosystem. Clearly, any objective of harvesting each commercial species at MSY makes no sense in the case of interdependent species (May et al. 1979).

The Antarctic Convention therefore requires (a) that any harvested population should not be decreased "below a level close to that which ensures the greatest annual increment"; (b) the "maintenance of ecological relationships between harvested, dependent, and related populations . . . and the restoration of depleted populations to the levels described in (a)"; and (c) "prevention of changes, or the minimization of the risk of changes in the marine ecosystem which are not potentially reversible over two to three decades." Vague as these requirements may be, they do at least constitute a departure from the MSY approach which has been traditional in international conventions.

Perhaps the most complex management scheme devised to date for a multispecies fishery was the Management Plan formulated by the U.S. Northeast Fisheries Management Council during the years 1977–1980. By 1980, total catch quotas had been established for over 20 species and were allocated as to area, season, and vessel class. Altogether, a list of over 600 regulations was in effect by 1980. With catches landed at more than 200 ports in New England, however, monitoring and enforcement of the regulations became obviously impossible (Wilson 1980, 1982). As a consequence, in 1982 the system was discarded and replaced with a simple set of regulations covering only the mesh size of nets. What the final management regime will be in this fishery is an interesting question.

It should be clear that no amount of bioeconomic modelling is going to "solve" the problems of multispecies fisheries. [See Mercer (1982) for a survey of multispecies approaches to fishery management.] The best that can be hoped for is that a useful classification of the issues, and perhaps some insights into appropriate management principles will emerge. With this modest aim we now proceed.

5.1   Mixed-Species Fisheries

By the term *mixed-species fishery* we shall mean a fishery in which more than one species is caught simultaneously by the gear. The typical example occurs in bottom trawling, where various species of demersal or groundfish are captured in the trawl nets.

An important characteristic of mixed-species fisheries is the likelihood that certain species may become depleted while others continue to support high catch rates. The general rule is that those species (or stocks) with comparatively low rates of intrinsic growth will tend to become progressively depleted in a mixed-species fishery, although the relative catchability also plays a role.

A simple model for the case of two species is (see Larkin 1966, Clark 1976a, chap. 9)

$$\left.\begin{array}{l} \dfrac{dX}{dt} = G_1(X) - q_X E X \\[2mm] \dfrac{dY}{dt} = G_2(Y) - q_Y E Y \end{array}\right\} \tag{5.1}$$

The two species, X and Y, are assumed to be ecologically independent, so that the natural growth function $G_1(X)$ is independent of Y, and vice versa. The same effort level E applies to both species, but the corresponding fishing mortalities $f_X = q_X E$ and $f_Y = q_Y E$ may be different, because of different catchability coefficients q_X, q_Y.

The possible equilibria of the above model ($dX/dt = dY/dt = 0$) are given by

$$\frac{G_1(X)}{q_X X} = \frac{G_2(Y)}{q_Y Y} = E \tag{5.2}$$

For example, if the growth functions are of logistic form, $G_1(X) = r_X X (1 - X/K_X)$ and so on, this becomes

$$\frac{r_X}{q_X}\left(1 - \frac{X}{K_X}\right) = \frac{r_Y}{q_Y}\left(1 - \frac{Y}{K_Y}\right) = E \tag{5.3}$$

To be explicit, assume now that

$$\frac{r_X}{q_X} > \frac{r_Y}{q_Y} \tag{5.4}$$

The equilibrium curve (5.3) (here a straight line) then takes the position shown in Figure 5.1. The curve originates, for $E = 0$, at the natural equilibrium position (K_X, K_Y). As effort increases, the equilibrium point (X, Y) moves downward as both species become more intensively exploited. Ultimately, at point P in the figure, the second species Y disappears completely, while X continues to provide the sustained catch rate $G_1(X)$. The coordinates of the point P are $(X_1, 0)$, where

$$X_1 = K_X\left(1 - \frac{r_Y/q_Y}{r_X/q_X}\right) \tag{5.5}$$

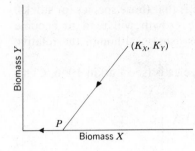

FIGURE 5.1 Equilibrium positions for the mixed-species fishery model. As effort E increases the equilibrium position moves down the broken curve, as indicated. At point P, species Y becomes (ultimately) extinct.

From Eq. (5.3) it follows that species Y will be eliminated provided

$$f_Y = q_Y E > r_Y \qquad (5.6)$$

that is, provided fishing mortality f_Y for species Y exceeds its intrinsic growth rate. Of course, the same holds for species X, but condition (5.4) guarantees that effort levels E exists at which (5.6) holds while

$$f_X < r_X$$

so that X persists while Y is eliminated. Thus the fishery may remain economically "viable" even though a valuable species, X, is ultimately fished out. In a sense, the introduction of man as a "superpredator" leads to a simplification of the ecosystem. The theory can be extended without difficulty to completely general models containing n species with arbitrary growth functions G. See Larkin and Gazey (1982) for a simulation-model approach.

The elimination of the less productive species is clearly possible under open-access conditions. Let π denote net revenue flow to the fishery and adopt the simplification of Chapter 1:

$$\pi = (p_X q_X X + p_Y q_Y Y - c)E \qquad (5.7)$$

Bionomic equilibrium is then characterized by the equation $\pi = 0$ together with the dynamic equilibrium equation, Eq. (5.3). The result is shown in Figure 5.2, in which two possible lines $\pi = 0$ are superimposed on the diagram of Figure 5.1. Since the line $\pi = 0$ intersects the X axis at $\bar{X} = c/p_X q_X$, we see that the condition under which species Y will be eliminated by the open-access (nonregulated) fishery is simply that $\bar{X} < X_1$, or

$$\frac{c}{p_X q_X} < K_X \left(1 - \frac{r_Y/q_Y}{r_X/q_X}\right) \qquad (5.8)$$

In other words, the low-growth-rate species Y will be eliminated pro-

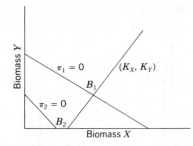

FIGURE 5.2 Bionomic equilibrium B in the mixed-species fishery for two positions of the curve $\pi = 0$.

vided only that species X is sufficiently valuable, regardless of the value of Y.

It is clear (without writing out any equations) that this principle extends to the case of n species: Those species with the lowest intrinsic growth rates may be eliminated or greatly reduced in number in an open-access mixed-species fishery. Such outcomes have often been noted in tropical trawl fisheries (Simpson 1982). However, the depletion of the larger species is often accompanied by increased stock levels of smaller, faster-growing species, presumably because of reduced predation or competition from the larger species. In the Antarctic, for example, the depletion of the largest whale species, blue and fin whales, was followed by an increase in the smaller sei and minke whales.

Another instance in which the principle probably applies is the haddock (*Melanogrammus aglefinus*) fishery in the northwest Atlantic. Annual landings of this species off New England, 1930–1978, are shown in Figure 5.3. According to Clark et al. (1980), haddock

supported New England's most important groundfish fishery for many decades and from the early 1920s to mid 1960s constituted one of the most important fishery resources in the United States.

The pronounced increase in landings in the late 1960s coincided with the entry of large distant-water mixed-species trawl fleets to the Georges Bank fishery. Clearly the haddock stocks were heavily overfished at this time. Although catches have recovered somewhat, they still remain well below historical levels. Most haddock are now taken as "incidental" catches in fisheries aimed at other groundfish species, and recovery of haddock may be inhibited as a result.

Declines in catches of the larger, more slowly growing (and often the most valuable) species in tropical trawl fisheries have been reported by many authors (Pauly and Murphy 1982). This phenomenon is doubtlessly a combination of the normal tendency in mature fisheries toward a

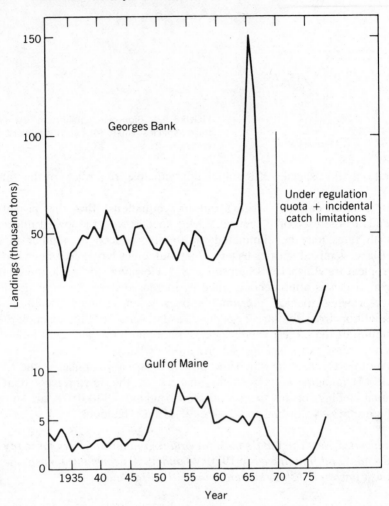

FIGURE 5.3   Annual landings of haddock in New England, 1930–1978. (From Clark et al. 1980.)

younger age composition in the catch and the principle of selective elimination discussed above.

Optimal Yield

The depletion of low-growth-rate species in a mixed-species fishery can obviously be prevented by reducing effort to an appropriate level. But this will obviously also imply a reduction in the catches of the other species involved—although in some cases it may be possible to employ selective fishing methods that will afford protection to critical species. For example, commercial fishermen in Lake Michigan are now constrained to using only stationary fish traps, from which lake trout, a favored sports-fishery species, must be released back into the lake (Talhelm 1978).

Various other methods, including gear restrictions and seasonal or area closures, can sometimes be used to control the depletion of vulnerable species in a mixed-species fishery. In some cases, retention of certain species taken as a by-catch may be prohibited, as has been the case with by-catches of Pacific halibut (*Hippoglossus hippoglossus*), salmon, and crab taken by groundfish trawlers in the northeast Pacific (Marasco and Terry 1982).

Since the discarded fish suffer virtually 100% mortality, such regulations may appear worthless, but they do of course affect the economic incentives of fishermen. Most fish populations have patchy distributions, and prohibition of one species will tend to influence the target areas selected by the fishermen, either before or during actual fishing operations (McKelvey 1981). Although some of the prohibited species may still be caught incidentally, the total catch should be reduced. However, enforcement of such regulations often proves difficult, especially in cases where partial processing of the catch is done at sea and species can be misidentified.

In many cases, however, effective selection of species or substocks may not be feasible in a mixed-species fishery. If not, the management agency faces the problem of either limiting the total catch so as to protect the vulnerable species or of allowing fishing on other species to proceed, "writing" off the vulnerable species. For the case of two species, optimum economic yield (zero discounting) is specified in our model by

$$\text{Maximize}_{E}\ (p_X q_X X + p_Y q_Y Y - c)E$$

$$\frac{dX}{dt} = G_1(X) - q_X XE = 0$$
$$\frac{dY}{dt} = G_2(Y) - q_Y YE = 0$$

Figure 5.4 shows the sustained revenue curves TR_X and TR_Y for the two species, in terms of effort E. As assumed earlier, species Y has the lower

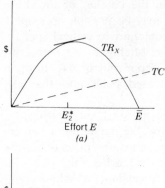

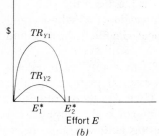

FIGURE 5.4 Optimum sustained yield in the mixed-species fishery: (a) species X; (b) species Y.

growth rate—see Eq. (5.4). Two different price levels $p_{Y1} > p_{Y2}$ for species Y are illustrated.

If Y is highly priced (p_{Y1}), the optimum effort E_1^* is largely determined by species Y, and X occurs primarily as a by-catch. In the figure, X is actually harvested at less than its MSY; species Y may be harvested either above or below its MSY effort level, depending on relationships between p_X, p_Y, and c. (In this case, fishermen might even be motivated to discard species X, even though $p_X > 0$—see below.)

If X is the highly priced species, on the other hand, the optimum effort E_2^* is entirely determined by the parameters for X. Here Y constitutes the by-catch but is progressively eliminated since E_2^* exceeds the maximum sustainable effort level for species Y. For example, it has been suggested that an optimal fishing strategy for North Sea groundfish might deliberately deplete cod, a "poor converter" of prey species, in order to obtain a larger yield of similarly valued haddock (McKellar 1982).

Note that, with the geometry of Figure 5.4, the open-access fishery eliminates species Y regardless of its price, besides leading to overfishing of species X. As usual, the optimum effort level E^* under positive discounting lies between the zero-discounting and open-access levels (Clark 1976a, p. 311).

The Pacific salmon (*Oncorhynchus* spp.) fisheries of British Columbia and the states of Alaska, Washington, and Oregon provide important instances of mixed-species and, more accurately, mixed-stock fisheries (Healey 1982). The different spawning stocks of salmon (even of the same species) are genetically distinct and should ideally be managed independently. (There are over 3000 separate stocks of salmon in British Columbia alone.) Unfortunately, with the exception of a few major runs, the salmon stocks cannot be separated at sea. Consequently many of the smaller, less productive stocks have become "overfished."

In an attempt to increase salmon production, the Canadian government has earmarked approximately $150 million for the construction of a series of hatcheries, artificial spawning channels, and other facilities under its Salmonid Enhancement Program. Unfortunately, various potential difficulties not originally taken into consideration have now been identified. For example, the artificially produced stocks of salmon will have above-average reproductive rates, since high production per spawner is a principal objective of enhancement facilities. Consequently, if the enhanced stocks are caught jointly with wild stocks, the latter (with their important genetic diversity) may ultimately disappear. This process seems already to have transpired in Washington State, where several natural salmon runs reached low levels following the introduction of enhancement facilities in the 1960s. In some cases this reduction may be the result of predatory or competitive interactions (Wright 1981). Other complexities and uncertainties also affect the enhancement problems.

It is clear that mixed-species fishery problems are complex and are becoming steadily more important. In many cases, the problems are exacerbated, if not actually caused, by poor management practices, particularly the failure to prevent overcapacity of fishing fleets. One problem that can become severe is that of discards, to which we now turn.

The Discard Problem

Species caught in a mixed-species fishery which have no commercial value are usually discarded at sea. In most cases the discarded catch will not even be recorded in the vessel's logbook.

But would fishermen ever have reason to discard valuable species? The answer is yes, for a variety of possible reasons:

1. Retention of the discarded species is prohibited.

2. The discarded species has a nonmarket value.

3. A valuable species is discarded to make room for more valuable species.

Examples of discarded species with nonmarket value are numerous. They include sports-fishery species, such as marlin taken by tuna long-liners, porpoises (*Stenella* spp. and others) which are valuable in helping Pacific tuna fishermen locate and capture tuna schools but which may be killed in the purse-seining process, immature fish of species of commercial value as adults, prey species, and so on. Preventing the incidental kill of such species may be difficult and not always worth attempting (Marasco and Terry 1982). It should be noted once again, however, that the problem may be exacerbated by management practices which force fishermen into a competitive scramble. Nonallocated TACs, closed seasons, and limited-entry programs in the customary sense all have this shortcoming.

Allocated quota systems may also lead to an increase in the discarding of valuable species, especially where quotas are nontransferable. In such a situation, any fisherman who fulfills his quota for a particular species is forced to discard the species while continuing to fish for other species.

It may seem unlikely that fishermen would deliberately discard marketable species, but the following simple model (Clark 1982) shows how this might come about. Let

$$h_j = q_j X_j E \tag{5.9}$$

denote the catch rate (by weight) of species X_j ($j = 1, 2, \ldots, n$) by a given vessel exerting effort E. Let T denote the total season length. For simplicity, we ignore depletion within a given season, so that X_j remains constant.

It t denotes the length of time spent fishing during one *trip* to the fishing ground, then $E = at$ is the total effort exerted for the trip ($a = $ constant), and hence we have

$$h_j = a_j t \tag{5.10}$$

where $a_j = aq_j X_j = $ constant. Now let θ_j denote the fraction of species X_j retained, $0 \le \theta_j \le 1$. The total retained catch for the trip is constrained by vessel capacity $\bar{H}$:

$$H = \sum_{j=1}^{n} \theta_j a_j t \le \bar{H} \tag{5.11}$$

Net revenue for the trip is given by

$$\pi = \sum_{j=1}^{n} p_j \theta_j a_j t - c_1 t - c_2 \tag{5.12}$$

where p_j are the prices, c_1 is variable effort cost, and c_2 is the fixed cost ("mobilization cost") for the trip. The n species will be listed in order of increasing price (by weight):

$$0 < p_1 < p_2 < \cdots < p_n \qquad (5.13)$$

[If two species happen to have the same price, they are simply lumped together and treated as a single species. Alternatively, the X_j could represent different size classes of a single species, having prices p_j dependent on size. Worthless species ($p = 0$) are ignored altogether.]

The fisherman wishes to maximize his net revenue for the season by making the optimal number of trips N and maximizing his revenue per trip. Since the model parameters all remain constant throughout the season, all trips will be of equal length $t = T/N$. First we treat t as fixed and simply maximize trip revenue π, as given by Eq. (5.12):

$$\underset{0 \le \theta_j \le 1}{\text{Maximize}} \sum_{j=1}^{n} \theta_j p_j a_j \qquad (5.14)$$

subject to

$$\sum_{j=1}^{n} \theta_j a_j \le \frac{\bar{H}}{t} \qquad (5.15)$$

(where the constant terms $-c_1 t - c_2$ have been dropped for simplicity).

The solution of this easy linear programming problem is intuitively obvious: Fill the hold with the more valuable species and discard the rest. Mathematically:

1. If

$$\sum_{j=1}^{n} a_j \le \frac{\bar{H}}{t}, \qquad \text{then } \theta_1 = \cdots = \theta_n = 1 \qquad (5.16)$$

(That is, for a short trip, all species are retained without exceeding the hold capacity.)

2. Otherwise, if

$$\sum_{j=k-1}^{n} a_j > \frac{\bar{H}}{t} \ge \sum_{j=k}^{n} a_j \qquad (5.17)$$

then $\theta_k = \theta_{k+1} = \cdots = \theta_n = 1$ and

$$\theta_{k-1} = \frac{\bar{H}/t - \sum\limits_{j=k}^{n} a_j}{a_{k-1}} \qquad (5.18)$$

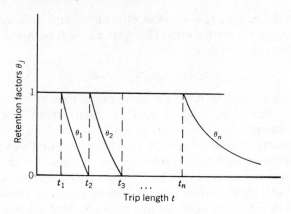

FIGURE 5.5 Retention factors θ_j for species X_j as functions of trip length t.

and $\theta_i = 0$ for $i < k - 1$ (retain all of species $X_k, \ldots, X_n$ and enough of X_{k-1} to fill the hold.) In particular, if the trip is very long, only the most valuable species X_n will be retained.

The values of the optimal $\theta_j s$, as functions of trip length t, are graphed in Figure 5.5; we have

$$\theta_k \begin{cases} = 1 & \text{for } t < t_k \\ \in (0, 1) & \text{for } t_k < t < t_{k+1} \\ = 0 & \text{for } t > t_{k+1} \end{cases}$$

where t_k denotes the length of trip such that the catch of species $X_k, \ldots, X_n$ completely fills the hold:

$$\sum_{j=k}^{n} a_j t_k = \bar{H}$$

It follows from Eq. (5.18) that the optimized net trip revenue, Eq. (5.12), is a piecewise linear convex function of t, with corners at the t_k and with (see Figure 5.6)

$$\frac{\partial \pi}{\partial t} = \sum_{j=k}^{n} (p_j - p_{k-1}) a_j - c_1 \qquad (t_{k-1} < t < t_k) \qquad (5.19)$$

We can now determine the optimal trip length. If N trips are taken during the season, total seasonal profit is given by $N\pi(T/N)$ or, writing $t = T/N$, by $T\pi(t)/t$. Thus the problem is simply

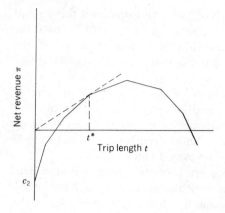

FIGURE 5.6 Optimized net trip revenue as a function of trip length t. The optimized trip length t^*.

$$\underset{t>0}{\text{Maximize}}\ \frac{\pi(t)}{t} \qquad (5.20)$$

In other words, the optimal trip length is that which maximizes average profit per trip. [A simple modification allows us to include "turnaround" time t_0—namely maximize $\pi(t)/(t+t_0)$.] The solution $t=t^*$ is indicated by the position of the dashed line in Figure 5.6; t^* in in fact a corner t_k of the curve $\pi(t)$.

(This is still only approximately correct, since $t=T/N$ can only assume discrete values; let us skip this technicality, which is minor and easily corrected by numerical calculation.)

The final outcome of the analysis is simply that discarding of valuable (but not the most valuable) species is optimal for the fisherman if fixed trip costs c_2 are large [or (exercise!) if the turnaround time t_0 is large]. The larger c_2 (or t_0), the longer the optimal trip length and the more species will be discarded; this is obvious from inspection of Figure 5.6.

What effect will quotas have on discarding behavior? This depends very much on the form of the quotas. One possibility would be a total allowable quota (by weight) on all species. If applied as a nonallocated TAC, such a quota would have no effect on individual trips [since the solution to problem (5.20) does not depend on $T=$ season length]. The season would simply close when the TAC had been achieved.

An allocated nonspecific quota, however, would motivate fishermen to fill as much as possible of the quota with valuable species and would thus lead to an increase in the amount of fish discarded (assuming that discards do not "count" toward the quota). While such a nonspecific quota system may be unusual for different species, the same problem can arise for quotas applied to a single species. Such allocated quotas can encourage the discarding of small fish of the given species if these small fish are less valuable than large fish. Separate quotas might be used for large and small fish, but this could easily make matters worse if the quotas failed to match the proportions in the catch. Again, transferability would be desirable.

It is also clear that, if a quota system of this kind leads to undesirable discarding practices, the distortion could be reduced by means of price adjustments. For example, high-priced species (or size classes) could be taxed, or low-priced species (or sizes) subsidized, or both, in order to remove the incentive to discard the low-priced catch.

Quotas applied to selected species are probably more common than general, overall quotas. Such quotas may also induce an increase in the amount of discarding. Suppose, for example, that a TAC is placed on the most valuable species X_n. Until the TAC is achieved, fishermen will tend to discard low-valued species, as before; once the TAC is achieved, X_n will be discarded (its price switches to zero), and lower-valued species will be retained instead. Unless the entire fishery is closed, the effect of the TAC may simply be to cause X_n to be discarded, although in practice fishermen will be motivated to target on other species.

The effect of an allocated seasonal quota Q_n on species X_n (for example) can be included in the above model by adding the constraint

$$\theta_n a_n < \frac{Q_n}{N}$$

to Eq. (5.15); here Q_n/N represents the quota per trip. (In the deterministic model we can assume that the fishermen divides his quota equally among the N trips.) As before, on each trip the fisherman fills his hold with the most valuable species but discards X_n when the trip quota Q_n/N is satisfied. Allocated quotas may thus lead to discarding of the regulated species in a mixed-species fishery.

Ecologically Related Species

The basic model used in this section, Eqs. (5.1), assumes that the two species X and Y are ecologically independent. A more general model is

$$\left.\begin{array}{l} \dfrac{dX}{dt} = G_1(X, Y) - q_X EX \\[12pt] \dfrac{dY}{dt} = G_2(X, Y) - q_Y EY \end{array}\right\} \qquad (5.21)$$

in which the natural growth terms $G_i(X, Y)$ explicitly involve dependence on both species. As before, the mixed-species aspect is reflected in the appearance of the same effort symbol E in the two equations.

A large variety of models of this type have been studied in the ecological literature (e.g., May 1975, 1980; Goh 1980). In the case that X is a predator with Y as prey, for example, one would assume that

$$\frac{\partial G_1}{\partial Y} > 0, \qquad \frac{\partial G_2}{\partial X} < 0$$

or, if X and Y were competitors,

$$\frac{\partial G_1}{\partial Y} < 0, \qquad \frac{\partial G_2}{\partial X} < 0$$

For constant E, the system (5.21) can be analyzed by traditional phase-plane methods. As shown by recent studies (Brauer and Soudack 1979), the dependence of solution trajectories (X_t, Y_t) on E can be quite complex and far from intuitive. However, these are general nonlinear phenomena, not peculiar to the mixed-species problems studied in this section. Further discussion is therefore postponed until Section 5.3.

Of course, mathematical models such as those in Eqs. (5.1) or (5.21) are ridiculously simplistic as representations of real marine ecosystems. Such models may be useful in providing qualitative insights into multi-species management problems (Section 5.3), but are obviously inadequate as a basis for quantitative application. The possibility of developing and using more realistic models will be discussed in Section 5.4.

5.2   Concentration and Switching

In Section 5.1 it was tacitly assumed that the·species X and Y of a mixed-species fishery were uniformly distributed over a single fishing ground—that is, that both species had type II concentration profiles. But as noted in Chapter 2, the assumption of uniform distributions can lead to serious bias and corresponding misunderstanding of fishery developments.

In the case of mixed-species fisheries, we will now show that when stock distributions are not uniform, fishermen may be motivated to undertake sudden shifts in "targetting" on different species or fishing areas. Such switching behavior, which is often observed, may of course also be related to informational considerations and possibly even to psychological aspects (Bockstael and Opaluch 1983). The switching behavior of Antarctic whaling fleets has been described and analyzed by Laws (1962). The following discussion follows Clark (1982).

Suppose there are n different fishing grounds, with stocks X_i and Y_i of two species, respectively. If total effort E_i is applied to the ith ground, the catch rates are

$$\left.\begin{aligned} h_X^i &= q_X^i E_i X_i \\ h_Y^i &= q_Y^i E_i Y_i \end{aligned}\right\} \tag{5.22}$$

where the catchability coefficients q_X^i, q_Y^i are assumed to be constant for each ground. The rate of return on ground i is given by

$$\pi_i = (p_X q_X^i X_i + p_Y q_Y^i Y_i - c_i) E_i \tag{5.23}$$

where c_i is the unit effort cost for ground i. Total effort capacity is denoted by E_{max}:

$$\sum_{i=1}^{n} E_i \leq E_{max} \tag{5.24}$$

As in Chapter 2, let us assume opportunistic fishing by omniscient fishermen. Thus effort will always be applied exclusively to that area (or those areas) with the largest return per unit of effort γ_i, where

$$\gamma_i = p_X q_X^i X_i + p_Y q_Y^i Y_i - c_i \tag{5.25}$$

In order not to introduce too many levels of complexity, let us ignore both the movement of fish between areas and all natural population processes. Also take $n = 2$. Then for $i = 1, 2$ we have

$$\left.\begin{aligned} \frac{dX_i}{dt} &= -q_X^i E_i X_i \\ \frac{dY_i}{dt} &= -q_Y^i E_i Y_i \end{aligned}\right\} \tag{5.26}$$

where

$$E_1 + E_2 = E_{max} \tag{5.27}$$

Suppose that initially we have $\gamma_1 > \gamma_2$, so that $E_1 = E_{max}$ and $E_2 = 0$. As X_1 and Y_1 are fished down, profitability γ_1 decreases. At some time $t_1 > 0$ we will have $\gamma_1 = \gamma_2$. For $t < t_1$ and t close to t_1 we must have either

$$q_X^1 X_1 > q_X^2 X_2 \quad \text{or} \quad q_Y^1 Y_1 > q_Y^2 Y_2 \tag{5.28}$$

but not both (although both could hold initially).

At $t = t_1$ some effort is switched from ground 1 to ground 2, and subsequently both grounds are exploited simultaneously, with $\gamma_1 \equiv \gamma_2$, that is,

$$p_X q_X^1 X_1 + p_Y q_Y^1 Y_1 \equiv p_X q_X^2 X_2 + p_Y q_Y^2 Y_2 \tag{5.29}$$

Differentiating this expression and substituting from Eqs. (5.26), we obtain

$$\alpha_1 E_1 = \alpha_2 E_2 \tag{5.30}$$

where

$$\alpha_i = p_X (q_X^i)^2 X_i + p_Y (q_Y^i)^2 Y_i \tag{5.31}$$

Hence, by Eq. (5.27),

$$E_1 = \frac{\alpha_2}{\alpha_1 + \alpha_2} E_{max}, \qquad E_2 = \frac{\alpha_1}{\alpha_1 + \alpha_2} E_{max} \tag{5.32}$$

Since the coefficient $\alpha_1/(\alpha_1 + \alpha_2)$ is obviously positive, we conclude that the distribution of effort between the two grounds shifts *discontinuously* at $t = t_1$. Likewise, aggregate catch per unit effort $(h_X^1 + h_X^2)/(E_1 + E_2)$ for species X, and similarly for Y, also shifts discontinuously at t_1. Prior to t_1, for example, aggregate CPUE for X is equal to $q_X^1 X_1$, whereas just after t_1 it becomes

$$\frac{q_X^1 X_1 E_1 + q_X^2 X_2 E_2}{E_1 + E_2} < q_X^1 X_1$$

by Eq. (5.28). Thus CPUE for X shifts downward; similarly, CPUE for Y shifts upward at t_1—see Figure 5.7.

CPUE on each ground separately, of course, does not switch discontinuously (and in fact is linear, by the assumption of constant q values on each ground). In practice, however, the switching of target areas by

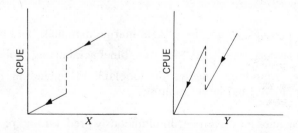

FIGURE 5.7 The switching effect for aggregate CPUE in a two-ground mixed-species fishery. (From Clark 1982.)

fishermen may not always be reflected in the catch-effort data available for analysis, so that biases in estimates of stock abundance can arise (Laws 1962).

The term *target species* is often encountered in the literature. Our present discussion, however, suggests that a more appropriate term in many cases would be *target area*, although species which are discarded could certainly be considered nontarget species. The same applies to the term *by-catch*, although again there are some valid uses of the word (to refer to the incidental catch of prohibited species which are therefore discarded, for example).

Failure to identify which species are "target" species is often claimed as a fundamental difficulty in apportioning effort data in mixed-species fisheries. Again, our discussion suggests that this is the wrong emphasis; a better approach would be to disaggregate effort by fishing area, using CPUE for each species in each area as an abundance index in that area. Where the data warrant it, this approach is sometimes used. It is also a standard technique used with randomized stratified sampling in research-vessel stock surveys (Doubleday and Rivard 1981).

The tendency of fishermen to target on high-profit fishing grounds might be utilized by managers to control the species composition in mixed-species fisheries, to a certain extent. For further study of this question we refer the reader to McKelvey (1981).

5.3   A Model of Two Trophic Levels

Marine ecosystems are extremely complex, particularly in comparison with terrestrial systems (Larkin 1978, Steele 1974). A single species may be predatory on several species, while serving as prey for other species. For example, Atlantic herring (*Clupea harengus*) have the following (Sindermann 1979):

1. *Predators*: Many species of fish, marine mammals, and birds
2. *Competitors*: Mackerel, alewives, blueback herring, pollock, shad
3. *Prey*: Numerous species of zooplankton, including copepods, amphipods, and larval molluscs.

Two given species may even be simultaneously predator and prey of each other at different life stages (Larkin 1978), an example being Atlantic cod (*Gadua morhua*), which feeds on capelin (*Mallotus villosus*), which in turn feeds on cod eggs. Through cannibalism, many species are also self-predatory.

As a consequence of this complexity, predicting the effects of exploitation of one species on the abundance of ecologically interdependent species is usually difficult and often impossible on the basis of existing data (e.g., Akenhead et al. 1982). Computer simulation models, which have sometimes been developed for this purpose (e.g., Laevastu et al. 1982, Ursin 1982), can provide useful insights—often simply by clarifying gaps in the understanding of the ecosystem under investigation—but they are expensive to develop and to run, depend on an extensive data base, and are usually highly specialized. General principles of resource bioeconomics seem more likely to emerge from the study of simpler analytic models which are more amenable to manipulation and study (see Section 5.4).

In this section we take on the modest task of modelling a simple two-species predator–prey system in which both predator and prey are subject to exploitation. In contrast to the previous section, we shall now assume that the predator and prey species are captured independently. Practical examples of such systems include fur seals and cod, baleen whales and krill, and salmon and herring. A general review of simple fishery models of this kind, but with a minimum of economic content, appears in May et al. (1979).

Consider first the following model, in which X denotes the prey biomass and Y the predator biomass:

$$\left.\begin{array}{l} \dfrac{dX_t}{dt} = r - m_X X_t - \alpha X_t Y_t - H_{Xt} = G_1(X_t, Y_t) - H_{Xt} \\[2mm] \dfrac{dY_t}{dt} = \lambda \alpha X_t Y_t - m_Y Y_t - H_{Yt} = G_2(X_t, Y_t) - H_{Yt} \end{array}\right\} \quad (5.33)$$

The symbols used here (except r) have the usual interpretations: H_{Xt} and H_{Yt} are the harvest rates of the two species. The term $\alpha X_t Y_t$ in the first equation represents the rate of consumption of prey by predators; this form assumes that the prey are uniformly distributed (type II concentration). In view of the apparent evolutionary advantages of schooling in reducing predation, this assumption may be unrealistic. Alternative forms for the predator–prey interaction terms, based on various assumptions concerning prey concentration and predator behavior, have been discussed in the ecological literature (e.g., Comins and Hassell 1979).

The corresponding term $\lambda \alpha X_t Y_t$ in the second equation represents the natural growth rate of the predator population; predation on X is thus assumed to be the only form of sustenance for Y. With $H_{Yt} = 0$, note that there is a critical prey population X_c given by

$$X_c = \frac{m_Y}{\lambda \alpha} \qquad (5.34)$$

with the property that (with no predator harvesting)

$$\frac{dY}{dt} \begin{cases} >0 & \text{if } X > X_c \\ <0 & \text{if } X < X_c \end{cases} \qquad (5.35)$$

In terms of prey harvesting, this implies that severe exploitation of the prey species X results in the extinction of the predator Y.

The (dimensionless) parameter λ in the term $\lambda \alpha X_t Y_t$ has an interpretation in terms of "transfer efficiency" of prey biomass into predator biomass.

The first of Eqs. (5.33) assumes, inter alia, that the prey species X has a constant rate of recruitment r, or in other words, that recruitment is density-independent. This is of course an extreme case, and other possibilities will be included later. The expression

$$K_X = \frac{r}{m_X} \qquad (5.36)$$

may be thought of as the "carrying capacity" for the prey species, since it is the equilibrium for the prey biomass in the case of zero predation and zero harvesting. We assume that

$$X_c < K_X \qquad (5.37)$$

for otherwise the model has no equilibrium with positive biomass of both predator and prey.

Equations (5.33) are convenient for detailed bioeconomic analysis and permit quite definitive conclusions to be drawn. The extent to which these conclusions are dependent on the particular form of the model will be considered later on.

Only certain combinations (X, Y) of prey and predator biomass permit sustained-yield harvesting of one or both species. Specifically, since this requires $\dot{X}_t = \dot{Y}_t = 0$ as will as H_{Xt}, H_{Yt} constant and non-negative, we see that the *sustained-yield set* Ω is given by

$$\Omega = \{(X, Y) \mid G_1(X, Y) \geq 0, \ G_2(X, Y) \geq 0\} \qquad (5.38)$$

(where, of course, $X, Y \geq 0$ is presumed). The set Ω for Eqs. (5.33), as shown in Figure 5.8, consists of an open region bounded by the curves $G_1 = 0$, $G_2 = 0$, together with the boundary of this region, plus a segment $0 \leq X \leq X_c$ on the X axis. This latter corresponds to sustainable prey populations with an extinct predator population.

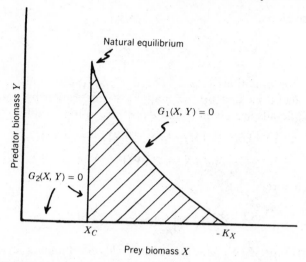

Prey biomass X

FIGURE 5.8   The sustained-yield set $\Omega(G_1 \geq 0, G_2 \geq 0)$ for Eqs. (5.33).

The natural equilibrium of the unexploited system has the coordinates

$$X = X_c, \qquad Y = \frac{1}{\alpha}\left(\frac{r}{X_c} - m_X\right)$$

A simple linearization analysis shows that this equilibrium is stable (although possibly oscillatory).

Let net fishery revenue for the two species be given by

$$\pi_X = [p_X - c_1(X)]H_X, \qquad \pi_Y = [p_Y - c_2(Y)]H_Y \qquad (5.39)$$

respectively; H_X and H_Y are assumed to be controllable independently. Simultaneous bionomic equilibrium in the two fisheries occurs when $\pi_X = \pi_Y = 0$; there are two cases to consider. Let $\bar{X}$ and $\bar{Y}$ be defined as usual by $p_X - c_1(\bar{X}) = p_Y - c_2(\bar{Y}) = 0$. If $\bar{X} > X_c$ (the critical prey population), then bionomic equilibrium occurs at $(\bar{X}, \bar{Y})$ in Ω. But if $\bar{X} < X_c$, the predator population is driven to extinction, and the ultimate equilibrium is at $(\bar{X}, 0)$, as noted earlier. In the former case, the marginal stock effect protects both species from extinction, but in the second case, the predators become extinct regardless of the marginal stock effect. It is interesting to note that the survival of the predator species is determined by a cost–price threshold for the prey species.

A sole owner of both fisheries would attempt to maximize the present value expression

$$\int_0^\infty e^{-\delta t}(\pi_{Xt} + \pi_{Yt}) \, dt \qquad (5.40)$$

subject to the equations (5.33), where H_{Xt}, $H_{Yt} \geq 0$. This optimization problem can be tackled by the method of integration by parts used in Chapter 1; details are given in the Appendix to this chapter. Define

$$V(X, Y) = [p_X - c_1(X)]G_1(X, Y) + [p_Y - c_2(Y)]G_2(X, Y)$$

$$- \delta[Z_1(X) + Z_2(Y)] \qquad (5.41)$$

where, as before,

$$Z_1(X) = \int_{\bar{X}}^X (p_X - c_1(X)) \, dX \qquad (5.42)$$

and similarly for $Z_2(Y)$. The optimal *equilibrium* biomass levels X^*, Y^* are then obtained by maximizing this function $V(X, Y)$ over the set Ω of feasible equilibria:

$$V(X^*, Y^*) = \max_\Omega V(X, Y) \qquad (5.43)$$

[It is much less obvious what the optimal *adjustment* policy from an initial position (X_0, Y_0) to the equilibrium (X^*, Y^*) is, but as argued in Clark (1976a, chap. 9), any "reasonable" method of reaching (X^*, Y^*) would probably be close to the optimum.]

It is clear that Eq. (5.43) does not depend on the particular functional form of $G_1(X, Y)$ and $G_2(X, Y)$ in the original model, Eqs. (5.33)—indeed, the same optimization criterion would apply to other types of interspecific relationships, such as competitive species. The validity of Eq. (5.43) does, however, require that the fisheries for X and Y be independent, that is, that the catch rates H_X, H_Y can be controlled separately. When this is not the case (as in a mixed-species fishery), it will generally not be possible to fish so as to achieve the equilibrium (X^*, Y^*). In fact, there is no analog to Eq. (5.43) for the mixed-species case, and no analytic solution to the present-value maximization problem is known. The difficulty is mathematically analogous to the mixed-cohort fishery model discussed in Section 3.7.

Let us return to the original model of this section, Eq. (5.33). Consider first the case in which fishing costs are negligible. Then, ignoring some irrelevant constant terms, we have

$$V(X, Y) = p_X[r - (m_X + \delta)X - \alpha XY] + p_Y[\lambda \alpha X - (m_Y + \delta)]Y \quad (5.44)$$

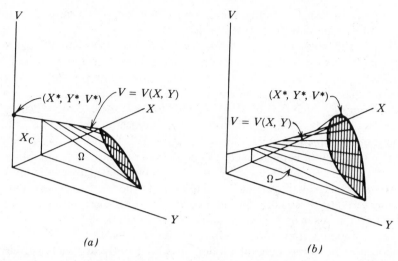

FIGURE 5.9 The value surface $V(X, Y)$; (a) highly priced prey species; optimum is MSY for prey and extinction of predators; (b) highly priced predator species; optimum involves exclusive predator harvesting.

The surface $V = V(X, Y)$ is thus a saddle surface, made up of straight-line cross sections in both the X and Y directions (Fig. 5.9). Clearly, V cannot have a maximum inside Ω, so the point (X^*, Y^*) must occur on the boundary of Ω. But this means that the optimal equilibrium involves *exclusive* harvesting, either of prey (Fig. 5.9a) or of predators (Fig. 5.9b). Which species is preferred depends, of course, on the relative prices p_X and p_Y (as well as on the other parameters).

If p_X is relatively large, for example, then prey are harvested exclusively in equilibrium, although of course predator harvesting would occur initially unless $p_Y = 0$. In fact, for the present model, the optimum is at $X^* = 0$ (and $Y^* = 0$), which is the "MSY" point for the prey species. (The limitations of the model in this respect were described earlier.) If p_Y is relatively large, on the other hand, the optimum is at some point (X^*, Y^*) on the curve $F(X, Y) = 0$; for $\delta = 0$ it can easily be seen that this is precisely the MSY point for the predator species.

What is the effect of discounting? It follows from Eq. (5.44) that $\partial V/\partial X$ and $\partial V/\partial Y$ are decreasing functions of δ; for δ sufficiently large (other parameters remaining fixed) we obtain $\partial V/\partial X < 0$ and $\partial V/\partial Y < 0$ throughout Ω. Thus, *regardless of the relative price levels*, maximization of present value at sufficiently high discount rates results in extinction of the predator population and exclusive harvesting of the prey. But this comes as no surprise, being merely an extension of the situation for single-species models described in Chapter 1.

In practice, such an extreme result has probably never actually occurred—nature is more complex than our models. However, the depletion of species at higher trophic levels and the subsequent development of fisheries based on species at lower levels appears to be a common phenomenon, which might be referred to as "trophic overfishing." Examples include whales and krill in the Antarctic and various trawl fisheries in tropical waters. As with the other forms of overfishing, trophic overfishing is most likely to occur under open-access conditions, but it could also arise as a result of discounting. However, unlike recruitment or growth overfishing, depletion of the predator species may be the (private or social) optimum, even for zero discounting. This is obvious in cases where the predators are economically worthless and merely compete with man for a share of the prey harvest.

Consider next the general model, in which costs are included; the functional forms of $G_1(X, Y)$ and $G_2(X, Y)$ are not specified. In order to stay within the confines of a predator–prey model, assume that

$$\frac{\partial G_1}{\partial Y} < 0 \quad \text{and} \quad \frac{\partial G_2}{\partial X} > 0 \tag{5.45}$$

The optimal equilibrium biomass levels X^*, Y^* are then characterized as before by the simple maximization problem

$$\underset{\Omega}{\text{Maximize }} V(X, Y) \tag{5.46}$$

where Ω and V are as before. Let $\bar{X}, \bar{Y}$ again denote the bionomic equilibria:

$$p_X - c_1(\bar{X}) = p_Y - c_2(\bar{Y}) = 0 \tag{5.47}$$

It is clear that at least one of X^*, Y^* must be greater than $\bar{X}, \bar{Y}$, respectively, for otherwise the sustained revenue from the two fisheries is negative. It is also intuitively obvious that X^* must be greater than $\bar{X}$, and this is easily proved to be correct [if $X^* \leq \bar{X}$, then $Y^* > \bar{Y}$, and a marginal increase in X^* increases the value of $V(X^*, Y^*)$, a contradiction.]

On the other hand, it is certainly possible that the optimal predator biomass Y^* may be smaller than the bionomic equilibrium $\bar{Y}$ in cases where the prey are more valuable than the predators. In the case that $X^* < X_c$ (the critical prey biomass) we will have $Y^* = 0$, and the predators will automatically be driven to extinction by the prey fishery (although predator harvesting may be appropriate during the initial development of the fishery). But even when $X^* > X_c$, it may happen that $Y^* < \bar{Y}$, in which case the predators should be harvested at a loss in order to increase the yield of the prey species. Under such circumstances, it may be appropriate to subsidize fishing on the predator since the individual fisherman will have no incentive to catch it.

Regulation

The relative simplicity of our present analysis of a predator–prey system is a consequence, as noted earlier, of the assumption of *separate* fisheries for the two species. This simplicity carries over to the analysis of the effects of regulation; all the difficulties of by-catches, discards, and so on disappear when the two fisheries are naturally distinct.

For example, it is straightforward to extend the analysis of Chapter 4 to demonstrate the equivalence of catch taxes and allocated transferable quotas and thus to show that either, or a combination of both, can be used to force the otherwise unregulated fishery to perform optimally. An obvious exception arises, however, in the situation just described, where predators should be harvested at a loss. Here we clearly obtain $\tau_Y < 0$—the "tax" on Y becomes a subsidy. But the quota market for predators is not likely to throw up a *negative* quota price—unless the fishermen are given quotas which they are *required* to catch (or to trade to someone who will). Alternatively, the prey quotas might require that a certain amount of predators be delivered along with the prey catch, thereby generating a demand for predators which would be equivalent to a subsidy. (A system similar to this was tried out briefly in the British Columbia salmon fishery in 1974, when salmon fishermen were required to turn in a certain amount of dogfish.)

To summarize, the regulatory methods which appear to offer advantages for the economic management of single-species fisheries, but which may encounter difficulties in the mixed-species setting, appear to remain effective in multispecies fisheries where the species àre harvested independently. These instruments might also be used in cases where some method can be found to transform a traditionally mixed-species fishery into a separated harvesting mode, provided the costs of doing so do not exceed the resulting benefits. The main problem in establishing appropriate quotas and/or taxes lies, as always, in the uncertainties inherent in the biological and economic systems.

5.4   Multicomponent Models

The two-species differential-equation models discussed in the preceding sections are little more than caricatures of the real complexity of marine food webs. At a somewhat more realistic level, Figure 5.10 is a diagrammatic representation of the principal energy flows among fish species in the North Sea (Jones 1982). But this model is also obviously a drastic simplification.

Of course, there is no difficulty in writing down larger systems of equations, which can also be modified to include age structure, random effects, and so on. Alternatively, complex flow diagrams can be developed and translated into computer simulation models.

The question is not whether such models can be constructed, but whether they will be of any value. At present I am unaware of any actual use of even a two-species model in fishery management. Computer simulation models, while popular in academic circles, have had little or no impact on management (Ursin 1982). This does not necessarily imply that such models are useless—they might well provide insights into marine ecology, leading ultimately to improvements in management. But there are valid scientific reasons not to expect too much progress in terms of quantitatively predictive multispecies models.

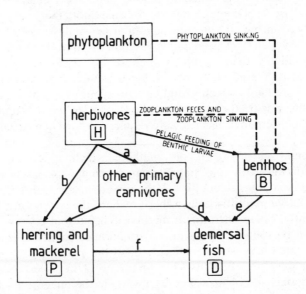

FIGURE 5.10   Diagrammatic representation of the principal energy flows among North Sea fish species. (From Jones 1982.)

First, such models are extremely data-intensive. Present fishery data barely supports the single-species (single- or multiple-cohort) models currently in use. For example, the most meticulous stock assessment programs seldom provide stock estimates with an accuracy exceeding ±50%. Moreover, time series are inevitably strongly affected by exogenous environmental factors which further screen the biological interactions. Attempts to estimate multispecies model parameters from such data are almost certainly doomed to failure.

A second limitation concerns the mathematical behavior of complex (even slightly complex) nonlinear models. Mathematical researches within the past two decades have unearthed whole new categories of unexpected behavior for such models, including bifurcations or "catastrophes" (Zeeman 1974), chaotic oscillations (May 1980), structural instability, and the occurrence of "strange attractors." Without going into details, it suffices to point out that these discoveries suggest that *the behavior of (deterministic) nonlinear system models is itself highly unpredictable*! Arbitrarily small changes in model structure or in parameters values can completely alter both the quantitative and qualitative long-term predictions of the model. (A mundane example would be: Develop a model to predict the shape of the smoke plume from a cigarette.)

Given that in fisheries systems parameter estimates with a standard error of ±50% are the norm and that model structures must be obtained largely by informed guessing, it is hardly surprising that progress has remained slow in terms of practical application of such models.

If it is true that realistic, predictive multispecies models are unattainable, how should management proceed? One suggestion is to continue using single-species models as a basis, introducing "adjustments" to catch quotas to take care of known species interactions (see Gulland and Garcia 1984). Another approach, which has been adopted in South African pelagic fisheries (Newman 1984), is to use a "total biomass" model in which all commercial species are simply lumped together. Because of the likelihood that the largest, slow-growing species tend to be the most valuable, and the first to be eliminated, this approach should clearly be used with discretion. For example, it should be possible to estimate a biomass-sensitive price function $p(X) = p_0 X^\beta$ $(0 < \beta < 1)$ from crude economic and biological data. The objective of maximizing price-weighted sustained yield might then be preferable, from an economic standpoint, to simply maximizing total biomass yield.

A less extreme approach might be to lump ecologically (or economically) similar species into a few main groups (Kirkwood 1982).

Multispecies management problems have been the subject of considerable recent discussion (May et al. 1979, Mercer 1982, Pauly and

Murphy 1982, May 1984). As in the case of single-species management, there is a strong tendency to ignore or trivialize economic considerations.

5.5 Summary and Perspective

It has often been asserted that single-species models are rapidly becoming obsolete as exploitation of fishery resources grows ever more intensive. It seems clear from the current literature, however (Mercer 1982, Pauly and Murphy 1982), that *operational* multispecies fishery models, capable of generating usable catch quotas derived from fishery or survey data, are not yet available. The quest to generate such models may well prove futile, owing to the irreducible complexity and uncertainty of marine ecosystems (Walters 1984).

In this chapter we have limited the discussion to simple multispecies models. The predictions, which are hardly surprising, include progressive depletion of slowly growing species in mixed-species fisheries, sudden switches of effort between different fishing areas as certain species are fished down, discarding of marketable species, and so forth. The effects of various regulations, including both allocated and nonallocated quotas, were discussed very briefly. Any regulatory system has the potential for introducing undesirable distortions, including increased discarding of valuable species.

Because of these and other complications, the management of multispecies fisheries will probably remain as much an art as a science. If so, the allocation of large budgets to resolve the "scientific" problems of multispecies fishery management may prove to be unwarranted in terms of the potential benefits.

Appendix

Dynamic optimization problems containing more than a single state variable are difficult to solve in general. Although the maximum principle as described in the Appendix to Chapter 2 applies in theory to such problems, there are many technical difficulties. In most cases, the two-point boundary-value problem generated by the maximum principle must be solved by numerical methods. Analogous problems arise in discrete-time optimization problems.

Detailed, technically correct solutions of dynamic multispecies optimization models would in any event appear to have limited practical value in terms of management policy, particularly in view of the current shortage of useful predictive models for marine ecosystems. Hence no attempt is made here to enter into theoretical technicalities. We will merely present some simple calculations to support the arguments given in the chapter.

Consider first the model of Section 5.3:

$$\text{Maximize}_{(H_{Xt}, H_{Yt})} \int_0^\infty e^{-\delta t}\{[p_X - c_1(X_t)]H_{Xt} + [p_Y - c_2(Y_t)]H_{Yt}\}\, dt \quad (5.48)$$

subject to

$$\left.\begin{aligned} \frac{dX_t}{dt} &= G_1(X_t, Y_t) - H_{Xt} \\ \frac{dY_t}{dt} &= G_2(X_t, Y_t) - H_{Yt} \end{aligned}\right\} \quad (5.49)$$

where the initial stock levels X_0, Y_0 are given, and

$$H_{Xt} \ge 0, \qquad H_{Yt} \ge 0 \quad (5.50)$$

As noted in the text, sustained-yield harvests are possible only for stock levels (X, Y) belonging to the set Ω given by

$$\Omega = \{(X, Y) \mid G_1(X, Y) \ge 0, G_2(X, Y) \ge 0\} \quad (5.51)$$

Since H_{Xt} and H_{Yt} are independent, we can substitute from Eqs. (5.49) into the integral of Eq. (5.48) and perform two integrations by parts, exactly as in Chapter 1. The resulting integral is

$$\int_0^\infty e^{-\delta t} V(X_t, Y_t)\, dt + Z_1(X_0) + Z_2(Y_0) \quad (5.52)$$

where

$$V(X, Y) = [p_X - c_1(X)]G_1(X, Y) + [p_Y - c_2(Y)]G_2(X, Y)$$
$$- \delta[Z_1(X) + Z_2(Y)] \quad (5.53)$$

with

$$Z_1(X) = \int_{\bar X}^X [p_X - c_1(u)]\, du \quad (5.54)$$

and so on. It follows that the point (X^*, Y^*) at which $V(X, Y)$ attains its maximum over the set Ω is an optimal equilibrium solution, in the sense that if $X = X^*$ and $Y = Y^*$, then no improvement in the present-value integral (5.52) can be achieved by changing X or Y. It is not obvious, however, what the dynamically optimal solution is from an initial stock level (X_0, Y_0) different from (X^*, Y^*). We will not take up this question here, but merely remark that any reasonable policy of adjusting stocks to (X^*, Y^*) should be approximately optimal (Clark 1976a, p. 324). For example, the single-species analog

$$H_{Xt} = \begin{cases} H_X^{\max} & \text{if } X_t > X^* \\ G_1(X^*, Y_t) & \text{if } X_t = X^* \\ 0 & \text{if } X_t < X^* \end{cases}$$

(and similarly for H_{Yt}) can be shown to be nonoptimal but should provide a near-optimal result.

6 Fluctuations and Uncertainty

It is widely recognized that fisheries science and management operate in a highly uncertain environment. Most fish stocks undergo wide fluctuations from one year to the next. Short-term predictions of stock abundance are generally inaccurate, and long-term predictions can only hope to identify trends. Since fish stocks cannot be observed directly, estimates of present and past stock levels are themselves subject to considerable error. Catch data are also often incomplete and unreliable—especially in cases where catch quotas have been imposed but not rigorously monitored. On the basis of such suspect data the fishery biologist is expected to estimate numerous parameters and thereby to determine optimal catch levels.

Many other sources of uncertainty affect fishermen, processors, marketers, and even consumers. The geographical location of fishable concentrations is often unknown, requiring extensive search operations. Prices and markets may be uncertain, and costs of fishing (fuel, financing charges, etc.) have also recently been subject to unusual degrees of uncertainty. The fish consumer may have difficulty assessing the quality,

207

freshness, and even the precise species of fish offered for retail sale. Finally, fishery managers may have difficulty predicting the effects of proposed regulations—and fishermen may have difficulty predicting changes in the regulations. (See Anderson 1984, Sissenwine 1984.)

How should these manifestations of uncertainty be incorporated into bioeconomic fishery models? What are the implications for fishery management? In particular, to what extent are deterministic models, which by definition ignore both random fluctuations and uncertainty, unsatisfactory for management purposes?

It is by no means easy to provide answers to these important questions. The existing literature in stochastic resource modelling is very recent and active but still somewhat disorganized. There is little in the way of accepted basic theory and no generally agreed-upon methodology. Standard practice consists usually of obtaining "best estimates" of parameters from available data and then utilizing these values in deterministic models, updating the estimates when new data become available (Walters 1981).

The simplest approach to explicit modelling of uncertainty is the direct "randomization" of a given deterministic model. Certain variables or parameters of the deterministic model are replaced with random variables having given probability distributions. Risk aversion may be included in the analysis by introducing a utility function $U(\pi)$ satisfying $U'(\pi) > 0$ and $U''(\pi) < 0$ (see Section 6.3). When applied to a dynamic model, this randomization leads to a so-called stochastic process and in the control setting to a problem in stochastic control theory.

For example, the equation

$$X_{t+1} = Z_t G(X_t - H_t) \qquad (6.1)$$

has been used to model population dynamics stochastically (e.g., see Reed 1979). The underlying deterministic model, given by $X_{t+1} = G(X_t - H_t)$, is modified by a multiplicative noise factor Z_t. The random variables Z_t, representing environmental disturbances, are assumed to be identically distributed, with probability density function $f(z)$ having mean $\bar{z} = 1$ and variance σ_z^2.

Given the current year's escapement $S_t = X_t - H_t$, Eq. (6.1) specifies the probability distribution for the population size next year; this distribution has mean $G(S_t)$ and variance $[\sigma_z G(S_t)]^2$—see Figure 6.1. An optimal harvesting model based on Eq. (6.1) will be discussed in the next section.

This randomization method provides a rather restricted approach to the problem of making decisions under uncertainty (Loasby 1976). In a sense, the randomization approach applies only to "gambler's uncertainty," wherein the probability of any possible outcome is assumed to be known

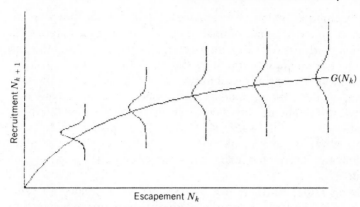

FIGURE 6.1 Probability distributions for the stochastic stock-recruitment model of Eq. (6.1). For each value of N_k, this distribution has mean $G(N_k)$ and variance proportional to $G(N_k)^2$.

and only the results of future "rolls of the dice" are unpredictable. Once the dice are rolled, the outcome is then assumed to be known exactly.

In fishery problems uncertainty is usually much more pervasive and fundamental than this. The main quantities of interest, such as population size and structure, are not directly observable and must be estimated from unreliable data; the appropriate probability distributions and their parameters are not known and must also be estimated from the data. The "best" model to use is itself seldom known.

Standard statistical procedures are of course available for the purpose of estimating both state variables and parameters from available fishery data (Ricker 1958, Gulland 1966). Ideally, such estimates should always be accompanied by analyses of their expected accuracy in terms of variances, confidence limits, and so on. It often turns out that the confidence limits obtained from fishery data are extremely wide.

This method—the approach of classical statistics—possesses a major shortcoming from the viewpoint of formulating management decisions. Namely, whereas confidence intervals and the like provide some insight into the expected accuracy (or, more to the point, the *uncertainty*) inherent in the estimates, no suggestion is provided as to how this information might be employed in formulating management decisions. Should catch quotas, as calculated on the basis of various estimates, be increased or decreased in the event these estimates are known to be highly uncertain? The general impression among fishery scientists seems to be that catch quotas should always be reduced in the presence of uncertainty, but the fishing industry often seems to take the opposite view!

A technique which has been developed for the purpose of making decisions in the presence of uncertainty goes under the name of (Bayesian) *decision analysis*. With this approach the analyst expresses the uncertainty in a given estimate in terms of probability—the so-called *prior* probability distribution. This prior distribution then becomes part of the information utilized in determining management policy. Various concepts of "optimality" under uncertainty have been formulated in the decision analysis literature (see Raiffa 1968, de Groot 1970, Arrow 1974, Heyman and Sobel 1984). In this book we will usually consider the objective of maximizing discounted expected economic yield, or utility, from the resource.

Going a step further, the analysis can be extended to estimate the expected increase in return that would result from *reducing* the level of uncertainty, which may involve delaying the final management decision until more data have been collected. Such an analysis could be used in principle, for example, to assess the advisability of setting conservative catch quotas where stock levels (or other variables) are uncertain. Decision analysis could also be used to assess the value of stock surveys (which are often the most expensive component of fishery management), as well as other research activities.

The methods of decision analysis have as yet found little actual application in fishery management, however. Many scientists are trained mainly in classical statistics and are unfamiliar with the Bayesian techniques used in decision analysis. The mathematical theory underlying decision analysis can be rather complex, especially in the dynamic setting required in resource modelling. Yet decision analysis seems to offer the only hope for making fully rational decisions in any field as uncertain and complex as fishery management.

A brief description of Bayesian decision analysis appears in Section 6.3, and some selected fishery applications are presented in Section 6.4. In Sections 6.1 and 6.2 we introduce several stochastic models, but do not use the Bayesian approach.

6.1   Stock Fluctuations

In this section we consider the problem of managing a fluctuating fish stock. The fluctuations are assumed to be random (i.e., unpredictable), having a known probability distribution. Also, the size of the stock at any given time is assumed to be known exactly; it is only the future stock size that is uncertain. As noted in the introduction, these assumptions are usually quite unrealistic; more realistic models require the use of decision-theoretic methods to be discussed later.

Descriptive Theory

The first question is of a purely descriptive nature: What will be the effects of harvesting a fluctuating fish stock? Will the fluctuations become more or less pronounced under harvesting? What about the chances of stock collapse? In short, questions of the *stability* of exploited fish stocks are of interest to the fishery manager. Harvest policies that maintain stability will generally be favored over those that induce instability.

We begin with an observation which is both trivial and profound: A sustained-yield policy, especially a policy that attempts to maintain the estimated maximum sustainable yield (MSY), is highly destabilizing. Theoretically, this is trivial—if annual catches are maintained at a fixed level, then several consecutive years of below-average recruitment may result in the collapse of the fishery.

Because of the widespread practice, at least in the past, of managing fisheries on the basis of MSY, the foregoing observation is far from trivial in a practical sense. The current trend away from MSY management seems to have been largely motivated by its inadequacy in the presence of natural stock fluctuations, although many other considerations are also relevant (Larkin 1977, Holt and Talbot 1978).

Indeed, it should be apparent that *sustained-yield* (that is, *constant-yield*) *policies are never optimal* in any reasonable sense when fish stocks are subject to fluctuation. It is ironic to see the term *optimal sustained yield* now appearing in legislation as a replacement for *maximum sustainable yield*, since nonsustainability is one of the main reasons for abandoning MSY.

If we consider the four quantities

$$X = \text{fish stock biomass}$$
$$E = \text{fishing effort}$$
$$C = \text{annual catch}$$
$$\pi = \text{net annual revenue}$$

it becomes apparent that any policy designed to stabilize one of these quantities is likely to destabilize the other three. For example, a constant-escapement policy, which provides maximum stability for the stock size X, clearly implies variations in E, C, and π. Likewise a constant-effort policy will result in variations in X, C, and π; and so on. The interplay among these types of variation can be studied by means of simulation experiments or by various analytical methods (see May et al. 1978, Reed 1983). Rather than pursuing this descriptive theory, we shall consider the problem from the optimization point of view.

Optimal Economic Yield

The degree to which fluctuations in catch or in economic yield are undesirable will depend primarily on the possibilities of *substitution*. In the limiting case of perfect substitutability, fishermen and vessels would be able to find ideal alternative employment when a particular fish stock was low, and consumers would find ideal substitutes for fish when supplies were low. Under these quite unrealistic assumptions, fluctuations in C, E, or π become unimportant, and an appropriate objective for management is simply the maximization of expected total discounted net revenue.

The following stochastic stock-recruitment model has been studied by Reed (1974, 1979):

$$X_{k+1} = Z_k G(S_k), \qquad k \geq 0 \tag{6.2}$$

$$S_k = X_k - H_k \tag{6.3}$$

$$0 \leq H_k \leq X_k \tag{6.4}$$

The symbols X_k, S_k, and H_k have the same interpretation as in Section 3.6, namely recruitment, escapement, and harvest in year k, respectively. In Eq. (6.2) the deterministic stock-recruitment function $G(S_k)$ is multiplied by a random factor Z_k. We shall assume that the random variables Z_k are independent and identically distributed (i.i.d.), with common distribution function $\Phi(z)$, where

$$\Phi(z)\, dz = \mathrm{pr}(z \leq Z_k \leq z + dz) \tag{6.5}$$

(The theory can be extended to cover the case in which the random variables Z_k are serially correlated by using the theory of Markov decision processes—see Sobel 1982 and Spulber 1982.)

Equation (6.2) models randomness in the stock-recruitment relationship in a specific but realistic manner. Fluctuations are proportional to the expected recruitment level $G(S_k)$—see Figure 6.1. We assume that $\Phi(z) = 0$ for $z \leq 0$, so that recruitment is necessarily positive (unless $S_k = 0$). The specific functional form of the distribution $\Phi(z)$ need not be specified at this stage. A lognormal distribution is often considered appropriate, particularly when the relationship between spawning and subsequent recruitment can be considered the outcome of a sequence of independent random events to which the central limit theorem can be applied.

We shall consider the objective of maximizing the expected present value of net economic yield:

$$V(X_0) = \max_{H_k} E\left\{ \sum_{k=0}^{\infty} \alpha^k \pi(X_k, H_k) \mid X_0 \right\} \qquad (6.6)$$

where, as in Chapter 3, Eq. (3.86),

$$\pi(X, H) = \int_{X-H}^{H} [p - c(x)] \, dx \qquad (6.7)$$

The mathematical expectation $E\{\ldots\}$ in Eq. (6.6) is taken with respect to the entire sequence of random variables $\{Z_k\}$. The annual harvests H_k are not assumed to be specified in advance but will be determined reactively in each subsequent year once the recruitment level X_k is known. Such a reactive policy is usually referred to as a "feedback" control policy. (In the present formulation we assume that the recruitment X_k is known *exactly* each year prior to the specification of the annual harvest quota H_k. This rather unrealistic assumption will be relaxed in Section 6.4.)

The expression $V(X_0)$ in Eq. (6.6), which represents the optimized expected discounted present value of the initial stock X_0, is usually called simply the *value function*. The fundamental equation satisfied by $V(X_0)$, known as the *dynamic programming equation*, is

$$V(X_0) = \max_{0 \le H_0 \le X_0} [\pi(X_0, H_0) + \alpha E_{Z_0}\{V(Z_0 G(X_0 - H_0))\}] \qquad (6.8)$$

This equation is easily derived: Assume first that some initial harvest H_0 is specified. Escapement is then $S_0 = X_0 - H_0$, and next year's recruitment will be

$$X_1 = Z_0 G(X_0 - H_0) \qquad (6.9)$$

The value next year is then $V(X_1)$, and the corresponding discounted expected value equals $\alpha E_{Z_0}\{V(X_1)\}$. Thus the total value of the resource, given H_0, is equal to the expression in square brackets in Eq. (6.8). Finally, $V(X_0)$ is obtained by maximizing with respect to H_0.

Let $F(X)$ denote an indefinite integral of $p - c(X)$, so that $\pi(X, H)$ can be written as

$$\pi(X, H) = F(X) - F(X - H) \qquad (6.10)$$

For example, we can specify

$$F(X) = \int_{\bar{X}}^{X} [p - c(x)] \, dx \qquad (6.11)$$

where $p = c(\bar{X})$, in which case $F(X)$ represents the "immediate harvest value" of the resource stock X (Reed 1979).

By means of the substitution $S_0 = X_0 - H_0$, we can rewrite Eq. (6.8) in the form

$$V(X_0) = F(X_0) + \max_{0 \le S_0 \le X_0} [\alpha E_Z\{V(ZG(S_0))\} - F(S_0)] \qquad (6.12)$$

The expression in square brackets represents the discounted expected increase in the value of the resource over the next year, given an escapement level S_0. The optical escapement thus maximizes this expected increase.

Now it can be shown, under quite general hypotheses (Reed 1979) that the value function $V(X_0)$ is increasing and convex. Hence the expression in square brackets in Eq. (6.12) is also convex. Let S^* denote the value of $S_0 \ge 0$ at which this function assumes its maximum value. A moment's consideration then leads to the conclusion that the optimal feasible escapement is simply the minimum of S^* and X_0. In other words, S^* is the optimal *target escapement*, and the optimal first-year harvest is given by

$$H_0 = \begin{cases} X_0 - S^* & \text{if } X_0 > S^* \\ 0 & \text{if } X_0 \le S^* \end{cases} \qquad (6.13)$$

Thus the optimal harvest policy has exactly the same form as for the deterministic model discussed in Section 3.6—although of course the stochastic and deterministic-equivalent target escapements are not necessarily equal.

This result is elegantly simple, but the validity of the constant-escapement rule turns out to depend critically on the structure of our model. The result breaks down if (1) a nonlinear utility function is used in the objective (6.6)—see below; or if (2) the recruitement levels X_k are not known exactly (see Section 6.4) or the stock-recruitment function $G(S)$ involves uncertain parameters (Section 6.4) or the resource stock possesses age-structure or other structures. Nevertheless, constant-escapement policies are currently used explicitly in managing the Pacific salmon fisheries and at least implicitly in many other fisheries. In most cases, the escapement target is calculated with the objective of maximizing expected biological yield.

Let D^* denote the optimal escapement for the deterministic model

$X_{k+1} = G(S_k)$ obtained by averaging over Z_k in Eq. (6.2)—see Section 3.6. The relationship between D^* and the optimal escapement S^* for the stochastic model has been investigated analytically by Reed (1979). In most cases it turns out that $S^* > D^*$, which agrees with one's intuition that a "safety allowance" ought to be made in response to unpredictable stock fluctuations.

The operational question then is: By what percentage should the deterministic escapement be increased? To answer this requires numerical solution of the dynamic programming equation (6.8). Such solutions have been obtained by Ludwig and Walters (1982), Charles (1983b), and Clark and Kirkwood (1984) using different methods. The result is that, even for quite large-scale recruitment fluctuations (coefficient of variation to 100%), the effect on escapement levels is minor (less than about 5%). Furthermore, the loss in expected value resulting from using the wrong (deterministic) escapement level is even less significant (less than 1%).

These results suggest that simply passing from a deterministic model to a related stochastic model is likely to have very little quantitative effect on the outcome of an optimization analysis. Such stochastic generalizations of deterministic models are commonly treated analytically in the literature, but comparative numerical studies are rare. Experience with several such studies, however, strongly supports our general conclusion.

On the other hand, the treatment of uncertainties, as distinct from fluctuations, can lead to results that are significantly different from the deterministic case—as we shall see.

Yield–Variance Trade-offs

As we have seen, a harvest policy that maximizes long-run economic or biological yield is necessarily a constant-escapement policy, under our model assumptions. Such a policy, however, may result in large annual fluctuations in the catch. In order to reduce this variation it is necessary to sacrifice average yield at least to some extent.

For example, let us consider the case where the demand for fish is not perfectly elastic. The manager may then wish to determine catch quotas so as to maximize discounted expected social utility

$$\pi_s(X, H) = U_s(H) - c(X)H \tag{6.14}$$

where $U_s(H)$ denotes the social utility function (Section 3.2):

$$U_s(H) = \int_0^H p(H)\,dH \tag{6.15}$$

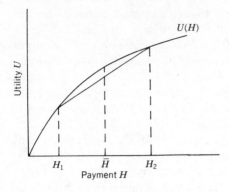

FIGURE 6.2 Risk aversion: $U(\bar{H}) >$ $\overline{U(H)} = [U(H_1) + U(H_2)]/2$.

Since $p(H)$ is assumed to be a decreasing function of H, we have

$$\frac{d^2 U_s}{dH^2} < 0 \qquad (6.16)$$

Under these circumstances it can be seen that variations in the annual catch reduce expected utility (see Fig. 6.2). This phenomenon is usually referred to as *risk aversion*, although perhaps the term *fluctuation aversion* would be more appropriate here.

Although the explicit solution of the maximization problem (6.2) based on the utility expression (6.15) does not appear to be known, the form of the solution can be conjectured from the deterministic theory (Section 3.2). The optimal harvest policy will be an *adaptive* (or "feedback") policy, that is, a policy of the form

$$H = H(X) \qquad (6.17)$$

which specifies the current catch H as a certain function of the current recruitment (stock) level X.

The standard harvest policies are all adaptive policies in this general sense (see Fig. 6.3):

Constant escapement: $H = \max(0, X - S)$

Constant mortality: $H = F \cdot X$

Constant catch: $H = Q = $ constant

but of course there are many other possibilities. For the case of nonlinear utility, the optimal harvest policy will presumably be an adaptive policy with a nonlinear function $H(X)$ of the form shown in Figure 6.4*a*. The nonlinear harvest policy permits small but positive catches if recruitment

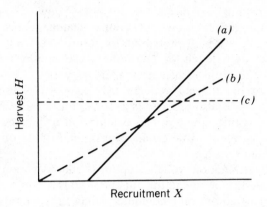

FIGURE 6.3 Three common harvest policies: (a) constant escapement; (b) constant effort (fishing mortality); (c) constant catch.

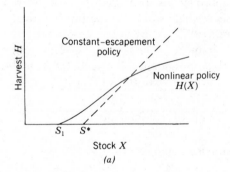

Constant-escapement policy

Nonlinear policy $H(X)$

Stock X

(a)

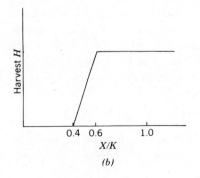

FIGURE 6.4 Optimal adaptive harvest policy $H(X)$ for nonlinear utility, compared with the constant-escapement policy: (a) general case; (b) IWC New Management Policy.

217

is below S^*—fishing is reduced, but not closed down completely unless recruitment is unusually low. Of course, catches must be reduced sufficiently to allow a high probability of future stock recovery. On the other hand, at high levels of recruitment the nonlinear policy yields smaller catches than the constant-escapement policy. Consequently, the annual variation in catches is reduced by means of this nonlinear policy. The average catch is also reduced (since the constant-escapement policy is known to maximize the average catch), and a detailed optimization analysis would attempt to determine the optimal trade-off between the average catch and the variance.

Explicit determination of the optimal adaptive policy, however, requires knowledge of the utility function $U(H)$, and the calculations must then be carried out on a computer, using the methods of stochastic dynamic programming. Utility functions themselves are difficult to specify and are subject to shifts over time. An alternative, more practical approach would involve simply an ad hoc selection of an adaptive policy using some reasonable adjustment of the optimal constant-escapement policy. As noted earlier, the latter can usually be obtained with sufficient accuracy from a deterministic model.

An example of the ad hoc approach is provided by the New Management Policy adopted for baleen whale stocks by the International Whaling Commission in 1979 (Birnie 1982). A given stock of whales is first classified as either *protected* (abundance less than $0.4K$, where $K =$ unexploited stock size), *managed* (abundance between $0.4K$ and $0.6K$), or *underexploited* (abundance greater than $0.6K$). The allowable annual catch C is then given by

$$
C = \begin{cases}
\bar{C} & \text{for underexploited stocks} \\[2mm]
\dfrac{X - 0.4K}{0.2K} \cdot \bar{C} & \text{for managed stocks} \\[2mm]
0 & \text{for protected stocks}
\end{cases}
$$

where $\bar{C} = 0.9 \times$ estimated maximum sustainable yield. This is illustrated in Figure 6.4b. This strongly conservative management policy was apparently adopted mainly on political grounds rather than as a policy designed to maximize any particular objective or to reduce variations. Whale populations are generally much less variable than fish stocks in any event; variations in an estimated stock of whales are likely due more to sampling errors (or to changes in the method of estimation) than to actual changes in the stock itself.

(Any adaptive management program will obviously face the problem of dealing with "phantom" fluctuations in the stock caused by inaccuracies in the estimates, but such forms of uncertainty are ignored in the present model—see Section 6.4.)

Other functional forms of adaptive policies can also be considered. Simulation exercises can then be performed to assess average yield and variance for such policies, but we shall not pursue the subject here. [An alternative approach to the yield–variance trade-off is discussed by Mendelssohn (1980a).]

Fishing Capacity

Another important aspect of stock fluctuations pertains to fishing capacity. The optimality result for constant-escapement policies assumes that there is no constraint on the size of annual catches. For a highly variable stock, a large capacity would be required in order to harvest and process the occasionally large allowable catch. When catches are at normal or subnormal levels, much of this capacity may be idled—unless profitable alternative uses are available, that is, unless the investment in capacity is *reversible* in the sense of Section 3.3.

The question of optimal investment in fishing capacity for fluctuating stocks has been studied by Charles (1983b), whose results can be summarized briefly as follows. First, for any existing level of capacity the optimal harvest policy is the following obvious modification of the constant-escapement policy:

$$H = \begin{cases} 0 & \text{if } X < S^* \\ \min(X - S^*, H_{max}) & \text{if } X > S^* \end{cases} \tag{6.18}$$

where H_{max} is the harvest obtained at full capacity.

Charles uses the standard model for within-season fishing (Section 3.6):

$$\frac{dx}{dt} = -qxE(t), \qquad x(0) = X \tag{6.19}$$

With maximum effort, $E(t) = E_{max}$, $0 \le t \le T$, we therefore have maximum catch given by

$$H_{max} = x(T) - x(0)$$

$$= [1 - \exp(-qE_{max}T)]X = \bar{F}_{max}X \tag{6.20}$$

where $\bar{F}_{max}$ denotes the whole-season fishing mortality. Note that (6.18) is an adaptive policy in our general sense (see Fig. 6.5). (Concentration profiles other than type II would result in a different shape for this curve.)

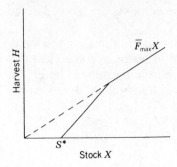

FIGURE 6.5 Constant-escapement policy with capacity constraint on effort: $H = 0$ for $X < S^*$; $H = X - S^*$ for $0 < X - S^* < \bar{F}_{max}X$; $H = \bar{F}_{max}X$ otherwise.

Secondly, Charles investigates the optimal investment strategy. New capacity can only be brought into the fishery with a one-year delay; a depreciation rate of 15% per annum is assumed. Figure 6.6 shows a typical result [using data from the northern Australian prawn fishery—Clark and Kirkwood (1979)]. The curve labelled $S(K)$ is the optimal escapement (which in fact varies slightly but inconsequentially with capacity K). The curve $h(S)$ specifies the optimal capacity (one year

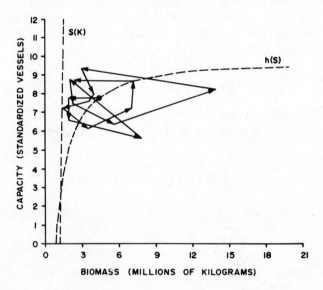

FIGURE 6.6 Control diagram for optimal investment and escapement, with stochastic recruitement. The curve $S(K)$ specifies the optimal escapement, given capacity K; the curve $h(S)$ specifies optimal investment given escapement S. The jagged curve is an optimized trajectory (15 years); note that investment occurs only in years when (S, K) lies below the curve $h(S)$. (From Charles 1983b.)

hence) as a function of current escapement biomass. [Because of the capacity constraint (6.20), this escapement is often higher than $S(K)$.] The jagged trajectory is the outcome of a 15-year simulation of the fishery; both escapement biomass and capacity fluctuate around the deterministic equilibrium.

Figure 6.6 constitutes a "feedback control diagram" analogous to the continuous-time deterministic model of Figure 3.7. Given the current recruited stock level X and existing capacity K, the optimal harvest H is given by Eq. (6.18). [In the simulated trajectory, in fact $H = H_{max}$ for 14 of the 15 years shown, and the escapement level $S(K)$ is only reached in one year: The capacity constraint usually determines the catch.] Then the optimal investment in new capacity for the coming year is given by

$$I^* = \max[h(S) - K, 0] \tag{6.21}$$

that is, capacity is built up to the level indicated by the curve $h(S)$. It can easily be verified that the trajectory shown does follow this investment rule. Note in particular that new investments are made following years of high escapement (resulting from high recruitment). This is because high escapement in one year implies a high probability of good recruitment the following year, and not because fishermen have lots of money to spend after a good season!

It may seem unreasonable, in Figure 6.6, that capacity should fluctuate in response to recruitment fluctuations. However, because the model assumes a 15% annual depreciation in capacity, this much replacement would be needed in any event in order to maintain a constant level of capacity. In the presence of natural fluctuations in the resource stock, it obviously makes sense to undertake these necessary investments in capacity when the resource is at a high level. If recruitment could be predicted ahead of time, the link between recruitment and investment would presumably become even stronger. (By the way, depreciation of fishing capacity might also be modelled as a random process, but the results of doing so now seem obvious.)

Charles (1983b) investigates the question of whether random fluctuations in the fish stock will cause optimal investment levels to increase or decrease relative to the deterministic model. The answer depends upon the balance between the "downside" risk of suffering idle capacity in bad years and the "upside" risk of lacking sufficient capacity to harvest in good years. When the cost of investing in capacity is low, a high level of "spare" capacity is desirable, and vice versa.

Finally, Charles discusses the sensitivity of the results to various biological and economic parameters and to the magnitude of the fluctuations. Except when the recruitment fluctuations are extremely large (coefficients of variation 100% or greater) it turns out that the policies calculated from a deterministic model perform almost as well as those obtained from the more difficult stochastic model. Thus, suppose that the policy curves $S(K)$ and $h(S)$ are derived via a deterministic model. These policies may differ substantially (by 30–40%) from the stochastically optimal policies, but the expected return obtained from the deterministic policies will differ to a smaller degree (typically less than 10%) from the stochastic optimum.

It seems to be emerging as a fairly general management principle that, if *properly interpreted*, management policies calculated from a deterministic model often remain appropriate under conditions involving random fluctuations. However, each case needs to be evaluated separately. There are exceptions to this somewhat vague principle, as our next model indicates. Also, the validity of the principle is certainly questionable in the case of true uncertainty, as contrasted to random fluctuations (see Section 6.4).

Opportunistic Fishing Fleets

The foregoing discussion of optimal capacity for fluctuating resources has been based on the assumption that fishing fleets are limited to exploiting a single fishery. In reality, many fishing fleets operate in an opportunistic fashion, switching their area of operation as the abundance of different fish stocks changes. The distant-water fleets of the Soviet Union, Japan, and other nations are capable of fishing virtually anywhere on earth. Tuna purse seiners from California exploit yellowfin and skipjack tuna in both the tropical Pacific and Atlantic oceans. Trawlers with freezing facilities can fish over wide ranges.

The advent of 200-mile Economic Zones has greatly reduced the freedom of such opportunistic fleets. The possibility exists, however, that opportunistic fleets could be used to advantage for fluctuating stocks, particularly where the coastal state has limited harvesting capacity. Rather than building up its own capacity to handle infrequent periods of abundance, the coastal state could consider selling the rights to surplus fish to foreign vessels on an occasional basis. Alternatively, if local harvesting capacity exists but processing capacity is limited, the state could allow sales of fish "over the side" to foreign factory vessels. Canadian fishermen, for example, have recently sold Pacific hake and Atlantic herring under such arrangements.

The economics of multipurpose, opportunistic fishing fleets operating in conjunction with single-purpose fleets has been modelled by McKelvey (1983) and by Beddington and Clark (1984). For the "local" fish stock X, both studies employ a stochastic stock-recruitment model, as in Eq. (6.2), but definitive analytical results are obtained only under the additional assumption that recruitment X_{k+1} is independent of previous escapement S_k. This "decoupling" of the stock-recruitment relationship greatly simplifies the analysis. As noted often in this book, the decoupling assumption is probably not unrealistic—at least on statistical grounds —provided escapement never falls to extremely low levels.

The results obtained by McKelvey and by Beddington and Clark can be described without going into the mathematical technicalities of their models. McKelvey's model pertains to the shrimp and groundfish fisheries of Oregon. The shrimp stocks are both more profitable and more variable than the groundfish stocks. Only the shrimp stock is modelled, therefore, with the groundfish stock being treated as an exogenous supply. Specialized shrimp vessels and generalized shrimp–groundfish trawlers exploit the shrimp stocks. The shrimp vessels are the more efficient in terms of variable costs of effort, but the groundfish fleets, which are always mobilized in the groundfish fishery, can switch cheaply and opportunistically when shrimp abundance is high.

With the objective of maximizing total revenue from the combined fleets, optimal harvesting involves the use of specialized vessels exclusively in the shrimp fishery when shrimp abundance is average or less, with the generalized trawlers entering only in years of high shrimp abundance. (Shrimp stocks vary from season to season over an order of magnitude; profitability is also influenced by fluctuations in shrimp prices, which McKelvey treats as exogenous.)

If the two fleets operate competitively under open access, the model predicts that the generalized fleet will come to dominate the specialized fleet, resulting in a suboptimal mix of the two fleets in the shrimp fishery. While questions of management policy are not addressed in McKelvey's study, it seems clear that an allocated quota system would require the use of *provisional* quotas for the generalized trawler vessels, such quotas becoming valid only in years of high shrimp abundance.

The model of Beddington and Clark (1984) is designed to analyze the problem of domestic vs. foreign harvesting of highly fluctuating coastal resources. (If the resource stock does not fluctuate significantly, it will usually be the case that the optimal fishery is either exclusively domestic or exclusively foreign. The same phenomenon arises in the McKelvey model.) Foreign fleets are charged a royalty on their catches, but the royalty they are prepared to pay is lower than the domestic value of the resource, so that the domestic fleets are favored. Because of the large fluctuations, however, it may not be wise to build up a domestic industry capable of handling the entire catch in years of high abundance. (Prices on domestic markets might also react to overproduction, but the model of Beddington and Clark does not treat this possibility.)

Both these models make the assumption that the annual recruitment R_k of the fluctuating resource stock will be fully known at the time when the catch allocation between specialized and generalized fleets is decided upon. (Recall that a similar hypothesis pertained to Reed's single-fleet model.) In practice, this assumption may be very far from correct. Estimates of R_k may be obtained from stock surveys, but these are usually very expensive and subject to considerably inaccuracy. An improved estimate may become available as the fishing season develops, but delaying the decision on the generalized fleet quota may not always be feasible.

But now we are beginning to consider the question of *uncertainty* in decision making; further discussion is therefore relegated to Section 6.4.

6.2   Other Types of Fluctuation

In Section 6.1 we dealt with random fluctuations in the size of the fish stock. Fluctuations in various other quantities are also important. The price of fish, for example, often responds to economic conditions exogenous to the fishery itself and hence can be treated as random. Other economic parameters, such as costs and interest rates, may also fluctuate, but it seems safe to assume that price fluctuations are generally most significant.

Fluctuations in the catch rates of individual fishing vessels are important in modelling the behavior of fishermen and the possible effects of regulations. In a deterministic framework we were able to establish the equivalence of individual catch quotas and taxes (royalties) (Section 4.3). It seems intuitively unlikely that this equivalence principle will persist in a stochastic framework; this nonequivalence will be established below.

Price Fluctuations

Consider first the case in which the price p_k of fish in year k fluctuates randomly with no discernible trend or serial correlation. The prices p_k are then independent, identically distributed random variables. Net revenue in year k is

$$\pi(X_k, H_k) = \int_{X_k - H_k}^{X_k} [p_k - c(x)] \, dx$$

For the objective of maximizing expected present value (i.e., fluctuations in revenue are not considered undesirable) we have

$$E_{(p_k)} \left\{ \sum \alpha^k \pi(X_k, H_k) \right\} = \sum \alpha^k \int_{X_k - H_k}^{X_k} [\bar{p} - c(x)] \, dx \qquad (6.22)$$

where $\bar{p} = E\{p_k\}$ is the average price. In this case, therefore, price fluctuations have no influence on management policy; one simply uses the average price $\bar{p}$ and proceeds as before. This applies whether the stock dynamics are treated deterministically or stochastically.

The above formulation assumes that the price p_k cannot be predicted prior to determining the annual catch quota H_k. As a more general model (which includes several interesting special cases), let us now suppose that the prices p_k are independent, but not identically distributed random variables, with known means $\bar{p}_k = E\{p_k\}$ depending on the time k. Equation (6.22) then becomes

$$E_{(p_k)} \left\{ \sum \alpha^k \pi(X_k, H_k) \right\} = \sum \alpha^k \int_{X_k - H_k}^{X_k} [\bar{p}_k - c(x)] \, dx \qquad (6.23)$$

Thus the optimization problem is again equivalent to a deterministic problem (assuming deterministic stock dynamics, for simplicity), but with time-varying price levels $\bar{p}_k$. The latter problem is myopic, as the usual calculation confirms:

$$\sum_0^\infty \alpha^k \int_{X_k - H_k}^{X_k} [\bar{p}_k - c(x)] \, dx = \sum_0^\infty \alpha^k \{\alpha[\bar{p}_{k+1} G(S_k) - C(G(S_k))]$$

$$- [\bar{p}_k S_k - C(S_k)]\} + \bar{p}_0 X_0 - C(X_0) \qquad (6.24)$$

where $C(x)$ denotes an indefinite integral of $c(x)$. The optimal escapement S_k^* in year k is the solution of

$$\underset{S_k}{\text{Maximize}} \, \{\alpha[\bar{p}_{k+1} G(S_k) - C(G(S_k))] - [\bar{p}_k S_k - C(S_k)]\} \qquad (6.25)$$

After differentiation, this yields the following rule, which will appear familiar [see Eq. (3.78)]:

$$G'(S_k^*) \frac{\bar{p}_{k+1} - c(G(S_k^*))}{\bar{p}_k - c(S_k^*)} = \frac{1}{\alpha} \qquad (6.26)$$

Thus when expected prices vary from one year to the next ($\bar{p}_{k+1} \neq \bar{p}_k$), the optimal escapement levels also vary. Note, however, that determination of S_k^* does not depend on *all* future price levels, but only on this year's and next year's expected prices. This is the sense in which a *myopic* decision rule is actually "myopic" [cf. Section 3.1, particularly Eq. (3.10)].

As a first application of this result, suppose now that the current year's price p is known precisely at the outset of the fishing season but that future prices are random with constant mean $\bar{p}$. Equation (6.26) then becomes

$$G'(S^*) \frac{\bar{p} - c(G(S^*))}{p - c(S^*)} = \frac{1}{\alpha} \tag{6.27}$$

Let $\bar{S}^*$ denote the optimal escapement at average future price levels:

$$G'(\bar{S}^*) \frac{\bar{p} - c(G(S^*))}{\bar{p} - c(S^*)} = \frac{1}{\alpha} \tag{6.28}$$

For example, if the current price is above average ($p > \bar{p}$), then Eqs. (6.27) and (6.28) imply that

$$G'(S^*) > G'(\bar{S}^*)$$

so that $S^* < \bar{S}^*$ [except for type IV concentration profiles, $c(x) = $ constant]. When prices are higher than usual, the stock should be exploited more intensively than usual—and vice versa, of course.

If variations in annual revenue are considered undesirable, this commonsense rule may no longer be valid. In fact, to reduce variations in revenue the opposite rule should be employed: *Reduce* catches when prices are high.

As a second application of the basic formula, Eq. (6.26), suppose fish prices are increasing, on the average, at a uniform rate:

$$\bar{p}_{k+1} = \lambda \bar{p}_k \qquad (\lambda > 1) \tag{6.29}$$

If marginal fishing costs are zero (type IV), Eq. (6.26) becomes

$$G'(S_k^*) = \frac{1}{\alpha \lambda} \tag{6.30}$$

A continuing increase in average price levels is thus equivalent to a decrease in the discount rate (i.e., to an increase in the discount factor α) and calls for a more conservative management policy than under constant price levels. When fishing costs are significant, the effect is modified but the general principle remains valid. Because of the myopic nature of Eq. (6.26), only short-term price predictions are needed.

[In extreme cases, the escapement levels S_k^* determined from Eq. (6.26) might vary more rapidly than the natural population growth, viz. $S_{k+1}^* > G(S_k^*)$. When this happens, the myopic solution is not feasible, and a "blocked interval" then arises. The situation is analogous to that discussed, in a deterministic continuous-time model, in Section 3.1.]

It should be clear that the foregoing analysis can be extended to include other sources of fluctuation. or example, the analogs to Eqs. (6.26) and (6.27) for the case where both price levels and stock levels fluctuate randomly are easily worked out; this question will be discussed further below. Similarly, fluctuations in discount rates can also be included.

Fishermen's Behavior

Next we consider the individual fisherman's reaction to price fluctuations (see Andersen 1982). We shall use the model of short-term fishermen behavior discussed in Section 4.2. Thus let E denote the total amount of fishing effort exerted by a given vessel over the "decision period"—say, one day, or perhaps, one fishing trip. If p denotes the price actually received for the catch, then the net realized revenue for the period is given by

$$\pi(E; p) = pqXE - C(E) \qquad (6.31)$$

Here we assume that q and X are fixed constants, the values of which are known to the fisherman (this restrictive assumption will be relaxed later).

Assume now that the fisherman is uncertain about the price that he will receive for his catch. How will this uncertainty affect his decision regarding effort E?

Naturally the fisherman has *some* prior expectation regarding price, derived, for example, from his knowledge of recent prices paid. The actual price paid may depend on the particular supply-and-demand situation at the time he delivers his catch; this is especially likely to be the case for fish supplied fresh directly to consumer or restaurant trade (Wilson 1980).

Let us suppose that the fisherman is sophisticated enough to construct a cumulative distribution $F(p)$:

$$F(p) = \text{pr}(\text{price} \le p) \qquad (6.32)$$

Thus $F(p)$ is an increasing function of p, with $F(p) = 1$ for sufficiently large p—see Figure 6.9 below. If there is a possibility that there will be *no* market for his catch, then $F(0) > 0$.

The effect of uncertainty on the fisherman's effort decision will depend on his degree of *risk aversion*. Namely, let $U(\pi)$ denote the fisherman's short-term utility function; we assume as usual that

$$U'(\pi) > 0, \qquad U''(\pi) \le 0$$

The case $U''(\pi) \equiv 0$ is called *risk neutrality* (this is discussed further in Section 6.3). We hypothesize that the fisherman will determine the period's effort E so as to maximize expected utility:

$$\max_{E \geq 0} E\{U(\pi(E, p))\} = \int U(\pi(E; p)) \, dF(p) \tag{6.33}$$

[The use of the same letter E with two distinct meanings in this equation should cause no confusion: the expectation operator $E\{\ldots\}$ is *always* accompanied by braces $\{\ldots\}$.]

The risk-neutral fisherman selects an effort level E_1 so as to maximize net revenue:

$$\underset{E_1 \geq 0}{\text{Maximize}} \ E_p\{pqXE_1 - c(E_1)\} = \bar{p}qXE_1 - c(E_1)$$

where $\bar{p}$ is the expected price to be received. Hence E_1 is determined by the "certainty-equivalent" rule:

$$c'(E_1) = \bar{p}qX \tag{6.34}$$

[or $E_1 = 0$ if $c'(0) > \bar{p}qX$].

The risk-averse fisherman determines his optimal effort level E_2 as in Eq. (6.33). It can easily be proved (see the Appendix to this chapter) that

$$c'(E_2) < \bar{p}qX$$

Under the assumption of increasing marginal costs, it therefore follows that $E_2 < E_1$ (except when $E_1 = E_2 = 0$); that is, the risk-averse fisherman employs a *lower* level of effort than does the risk-neutral fisherman. This result probably agrees with one's intuition.

The prediction, however, is a purely qualitative one. Without a knowledge of the individual fisherman's *short-term* utility function U and marginal cost function $C'(E)$ we cannot say by how much E_2 will differ from E_1. It seems unlikely that this utility function would be strongly nonlinear, since only the income from a single "effort period" (e.g., one trip) is involved. The difference between E_1 and E_2 also depends on the amount of uncertainty involved, that is, on the variance of p. But again it is only uncertainty pertaining to the current decision period that is relevant here. In most cases this uncertainty would be small, and often it is zero—for example, in situations where the price is negotiated in advance. In general, it therefore appears unlikely that uncertainty caused by fluctuations in the price level would have a very severe effect on the day-to-day operation of fishing vessels.

Long-term investment decisions in vessel or processing capacity, on the other hand, would be much more likely to be influenced by uncertainty regarding future market prices. The risk-averse investor will tend to invest less than the risk-neutral investor, depending on the level of uncertainty.

Some economists (see, e.g., Arrow 1974) have argued that society as a whole should be risk-neutral, although the subject remains somewhat controversial. If so, the possibility arises of *undercapitalization* of the open-access fishing industry. No doubt many programs of government subsidization of plant and vessel construction have been justified on the basis of such considerations. The problem is complex, since in fact uncertainty is also a dynamic quantity—as the fishery develops, uncertainty about markets and prices will tend to decrease. Rigorous analysis would thus require a decision-theoretic framework, as outlined in Section 6.3.

Fluctuations in annual income have been noted to have a strong effect on fishermen's investment behavior. When prices and catches both reach a peak in the same year, all fishermen may enjoy an income bonanza, thereby providing the opportunity to buy that new boat at last or otherwise to invest in increased fishing capacity.

For example, British Columbia salmon fishermen in 1973 were favored with unusually large runs of salmon, while at the same time the Alaskan runs were below nowmal. Combined with a general "commodity boom" in the same year, this resulted in unusually high salmon prices. The ultimate result was a substantial increase in the catching power of the B.C. salmon fleet, which in effect defeated the purpose of the salmon license-limitation scheme that had been inaugurated in 1969 (Fraser 1978).

If price p and catch C fluctuate independently with coefficients of variation CV_p and CV_c, respectively, then the coefficient of variation of gross income $I = pC$ is easily seen to be given by

$$CV_I^2 = CV_p^2 + CV_c^2 + CV_p^2 CV_c^2$$

Thus variations in gross income are larger than the individual variations in price or catch. Under a constant-escapement policy (as in the B.C. salmon fishery), variations in the total catch are themselves larger than the variations in stock levels. And variations in individual catches are likely to be even larger (see below). Fishing is indeed a risky profession!

The question of price fluctuations will not be pursued further here. The reader is referred to the paper of Andersen (1982) for further qualitative analysis of the effect of short-term price fluctuations on fishermen's behavior and the implications for fishery regulation.

Catch Fluctuations

In the above analysis of price fluctuations we have once again used the basic Schaefer expression

$$C = qXE \qquad (6.35)$$

to model, for example, daily catch C as a function of the daily effort E exerted by a single vessel. However, our more fundamental relationship between catch and effort (Section 2.2) is

$$C = \theta\rho E \qquad (6.36)$$

where ρ denotes the density of fish encountered by the gear and θ is a scaling constant.

In reality, this density ρ will generally be subject to short-term random fluctuations resulting from the patchy distribution of fish in the sea. For simplicity write $w = \theta\rho$; we now consider w as a random variable with known distribution $\phi(w)$. For type II concentration we then have

$$\bar{w} = E(w) = qX \quad (q = \text{constant})$$

For other concentration profiles, q will depend on X (see Section 2.2). In any case, for short-term effort decisions, $\bar{w}$ is a constant which we assume (unrealistically, of course!) to be known to the fisherman.

The mathematical analysis is now exactly parallel to the case of random price fluctuations discussed above; the fisherman chooses E so as to maximize expected utility:

$$\underset{E \geq 0}{\text{Maximize }} E_w\{U(pwE - c(E))\} \qquad (6.37)$$

We conclude, as before, that the risk-averse fisherman will exert less effort, because his daily catch is random, than will the risk-neutral fisherman. In theory at least, this could result in underexploitation of the open-access fishery.

To what extent is risk aversion likely to have an actual effect on fishermen's behavior? Fluctuations in the daily catch are surely a way of life for most commercial fishermen. There seems to be little reason to expect fishermen to be particularly averse to *short-term* fluctuations in their income; indeed, many fishermen may have a gambler's taste for the daily ups and downs of their vocation. Such risk-favoring fishermen would tend to exert *more* fishing effort than the risk-neutral fisherman.

For longer-term decisions, however, risk aversion is much more likely to become significant. Most people who happily gamble with a day's wages would hesitate to gamble with their annual income. Although we shall not attempt to model long-term investment decisions under uncertainty here, the undercapitalization effect of long-term risk aversion noted above for the case of price uncertainty is likely to become more pronounced when other forms of long-term uncertainty, such as uncertainty pertaining to future stock abundance, are also taken into account.

The overall prediction of this rather diffuse theorizing is that in certain situations the open-access fishery may be observed to suffer simultaneously from overfishing and undercapacity. Such a condition might arise if fishermen are favorable (or neutral) toward small, short-term income fluctuations but risk-averse where large-scale investments are concerned. Because of the risky nature of their profession, fishermen also often have little access to capital markets. While vessel subsidization may well be justified under these circumstances, the general trend toward overcapacity, rather than the reverse, must be kept in mind. Where government fishery policy combines measures to prevent overfishing with subsidization of capital costs but fails to resolve the common-property situation, vast overcapitalization of the fishery can hardly be avoided.

Fishermen's Quotas

The use of transferable fishermen's (or vessel) quotas as a method of counteracting the competitive, common-property externality in fisheries was discussed in Section 4.3, on the basis of a deterministic model. Attempts to introduce such quota systems have often met with at least initial resistance on the part of the fishermen themselves, for reasons which are not always immediately clear. One possible explanation, which we shall now investigate (Clark 1984), is related to the variation in catches which fishermen may be accustomed to.

The fisherman whose prospects are dominated by the chance of an occasional large catch may see a personal quota as a serious constraint. No longer will it be possible, with the quota in effect, to take advantage of a lucky catch. If a market develops for the transfer of quotas, then the bonanza of a lucky catch may still be realized—but with benefits reduced by the cost of purchasing additional quota units. (From the management viewpoint, these quota prices are eminently desirable, as explained in Chapter 4.) On the other hand, in unlucky years the fisherman's income will be increased if he is able to sell off unused quota units. The individual quota system thus has the potential—*assuming transferability*—of stabilizing fishermen's incomes.

The following simple model (Mangel and Plant 1984, Clark 1984) describes the situation. Let $p(n, \alpha)$ denote the probability of catching n

tonnes of fish ($n = 0, 1, 2, \ldots$) during a given period of fishing; here α is a vector parameter which depends on the particular probability model selected. If a nontransferable individual quota of Q tonnes per fishing period is imposed, then the probability distribution $p(n, \alpha)$ is truncated:

$$
p(n; \alpha, Q) = \begin{cases} \dfrac{p(n, \alpha)}{b_Q} & \text{if } n < Q \\ 0 & \text{otherwise} \end{cases} \tag{6.38}
$$

where the normalization factor b_Q is given by

$$
b_Q = \sum_{n=0}^{Q} p(n, \alpha) \tag{6.39}
$$

(For simplicity we treat n as a discrete variable here; the continuous case is analogous.)

The fisherman's expected catch over the specified period, with quota Q, is given by

$$
E\{n \mid Q\} = \sum_{n=0}^{Q} np(n; \alpha, Q) \tag{6.40}
$$

It is intuitively clear (and not too hard to prove) that

$$
E\{n \mid Q\} < E\{n\} = \sum_{n=0}^{\infty} np(n, \alpha)
$$

that is, the existence of a quota decreases the expected catch. However, the extent of the reduction can only be assessed after an appropriate distribution $p(n, \alpha)$ has been specified.

In the case that fish are "uniformly" randomly distributed over the fishing ground, the Poisson model is appropriate (see also Section 2.4):

$$
p(n, \alpha) = p(n, \lambda) = \frac{\lambda^n}{n!} e^{-\lambda} \tag{6.41}
$$

The mean $\bar{n}$ and variance σ_n^2 of the Poisson distribution are both equal to the search parameter λ, which is proportional to the search effort or fishing effort expended during the period in question.

A more general model can be obtained by supposing that fish are grouped into "clumps" of variable size, the clumps themselves being distributed according to the Poisson model. The resulting distribution then depends upon the distribution of clump sizes. In the event that the latter is a logarithmic distribution, it can be shown that $p(n, \alpha)$ is a negative binomial distribution (Pielou 1977, p. 118):

$$p(n, \alpha) = p(n, x, k) = \frac{\Gamma(n+k)}{k^n \Gamma(k)} \frac{x^n}{n!} \left(1 + \frac{x}{k}\right)^{-n-k} \qquad (6.42)$$

The parameters x and k are both directly proportional to the search parameter λ of the underlying Poisson model, with coefficients determined by the parameter of the logarithmic clump-size distribution. The negative binomial distribution is frequently used to model clumped distributions of biological organisms (Pielou 1977).

The mean and variance of the negative binomial distribution are given by

$$\bar{n} = x, \qquad \sigma_n^2 = x + \frac{x^2}{k} \qquad (6.43)$$

Thus the variance is larger than the mean, by an amount that varies inversely with the "contagion parameter" k. As $k \to \infty$, it can be seen that the negative binomial distribution $p(n, x, k)$ approaches the Poisson $p(n, x)$.

The values of x and k can be estimated from catch-per-vessel data according to the formulas

$$x = \bar{n}, \qquad k = \frac{\bar{n}^2}{\hat{\sigma}_n^2 - \bar{n}} \qquad (6.44)$$

As noted above, both x and k are directly proportional to λ, the search parameter. This implies that data from different vessel types and over different search periods can be combined and used in estimating the parameters x and k, provided the relative search power of different vessels has been calibrated. [There is a statistical difficulty, however. Given that the distribution of fish is clumped, we can expect that the distribution of fishing effort will also tend to be clumped! Hence the Poisson search model may be a poor choice, although it is not clear what model should be used instead. See Leaman (1981) for further discussion and references.]

Figure 6.7 shows the expected catch $\bar{Q} = E\{n \mid Q\}$ as a function of Q, for the negative binomial model, for various values of the contagion parameter k. (In this figure we have taken $Q = x$, the expected catch rate with no quota constraint.) The reduction in expected catch is high when k is small, and vice versa; for example, $k = 0.1$ leads to about a 91% reduction in expected catch: $\bar{Q} = 0.09 Q$.

Thus if individual quotas are imposed in a fishery based on a highly patchy fish stock (as evidenced by large time variation in individual catches), then it can be anticipated that fishermen's average catches will be significantly reduced because of the quota restriction. Pooling of

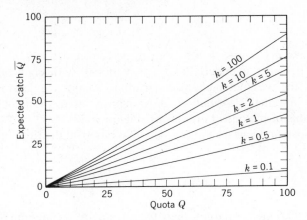

FIGURE 6.7 Expected catch $\bar{Q}$ vs. quota Q for various values of the contagion parameter k. (From Clark 1984.)

quotas, however, could reduce the effect. As an example, suppose that $k = 1.0$ and that 5 fishermen each have a quota of $Q = 20$ tonnes/week. According to Figure 6.7, the individual's expected catch is 8 tonnes, that is, a reduction of 60%. If the 5 quotas are pooled to yield $Q_{total} = 100$ tonnes/week, the expected total catch becomes 68 tonnes, a reduction of only 32%. (Note that the curve for $k = 5.0$ must be used for the pooled quota, since the search rate for 5 vessels is 5 times the individual rate.) Similarly, 10 fishermen pooling their quotas could expect to land 76% of their pooled quota. Of course, this assumes that fishermen will work just as hard to catch fish for this pool as for themselves—perhaps a dubious assumption.

The same calculation applies to the pooling of quotas over *time*. For example, if $k = 1.0$, a fisherman with a quota of 20 tonnes/week expects to land 8 tonnes/week on the average. But if the quota is administered instead as a 100-tonne quota over a 5-week fishing season, the expected total catch is 68 tonnes, by the same reasoning as before (the Poisson catch parameter is multiplied by 5, implying $k = 5.0$).

For some reason, the pooling or transfer of quotas often seems to be viewed with alarm by fishery administrators. Possibly it is felt that monitoring and enforcement would become difficult if pooling were permitted. Our simple model suggests, however, that pooling would be desirable from the efficiency viewpoint, especially in the case of patchy fish stocks. Also, the time period to which quotas apply needs to be carefully considered. Other schemes than those modelled here could be used—for example, carryover of unfilled quotas from one week to later weeks would help the fisherman who encounters bad luck in the early days of the season.

A potential problem with quota pooling, alluded to above, concerns the degree of mutual trust between members of the pool. An alternative to the pooling would be the establishment of a market for unfillled portions of individual quotas. The dynamics of such a market might be rather complex and could somewhat resemble the market for stock options. The relationship between taxes and quota market prices (cf. Section 4.3) might be clarified in the stochastic setting by modelling the quota market, but we shall not pursue the study here (see Weitzman 1974).

6.3   Elementary Decision Theory

In this section we present a very brief and superficial introduction to the theory of decision making under uncertainty. The reader is referred to the extensive literature for further details (Raiffa 1968, Schlaifer 1969, de Groot 1970, Arrow 1974). This literature does not pay much attention to dynamic decision problems of the type encountered in fishery modelling, however.

Insurable Risk

The simplest situation is exemplified by the decision problems facing a gambler playing an artificial game, such as roulette, or any of the numerous card and dice games. Such decision problems are characterized by *complete knowledge of the probabilities involved*. For example, the probability of rolling any particular number with two "fair" dice can immediately be calculated from the fact that each separate face turns up with probability 1/6.

Let p_n denote the probability of outcome n, and let $w(n, a)$ be the payoff received by the player when outcome n occurs and when the action denoted by a has been taken. Then the *expected payoff* from a play of the game is defined as

$$E_n\{w(n, a)\} = \sum_n p_n w(n, a) \qquad (6.45)$$

The simplest stochastic decision rule is then: Choose action a so as to maximize this expected payoff. Successful dice, poker, and bridge players all commonly employ this rule, which may of course involve a great deal of skill in execution.

In the case of the insurance industry, the probabilities of fire, accident, death, and so forth are obtained empirically and listed in appropriate actuarial tables. Risks which are or could be covered under such policies are referred to as *insurable risks*: Such risks are characterized by having fairly accurately *known probabilities*. (It is sometimes argued that these probabilities may be altered by insurance, a phenomenon referred to as "moral hazard." Although this phenomenon may well arise also in the fishery setting, we shall not be concerned with it here.)

The gambler betting on a single roll of the dice, like the insurance firm issuing a homeowner policy, is usually dealing with a payoff which is small relative to total assets. Under such circumstances the objective of maximizing expected values, as in Eq. (6.45), is reasonable. The expected-value objective weighs losses and gains equally; the gambler is said to be "risk-neutral" with respect to such wagers.

In cases where the payoff represents a significant fraction of one's total assets, however, the assumption of risk neutrality becomes questionable. Instead it is usually postulated that the gambler (or decision maker) values payoffs w in terms of a nonlinear *utility function* $U(w)$, satisfying the conditions

$$U'(w) > 0, \qquad U''(w) < 0 \tag{6.46}$$

as illustrated in Figure 6.8.

The *expected utility* of a gamble $w(n, a)$ is given by

$$E_n\{U(w(n, a))\} = \sum_n p_n U(w(n, a)) \tag{6.47}$$

For example, imagine a 50–50 bet ("heads or tails") with payoff $\pm\$100,000$. We have $p_1 = p_2 = \frac{1}{2}$ and $w_1 = -w_2 = \$100,000$. The expected payoff (6.45) is zero, but the expected utility is

$$E\{U(w)\} = \frac{1}{2} U(w_1) + \frac{1}{2} U(w_2) < 0$$

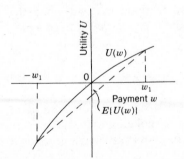

FIGURE 6.8 Utility $U(w)$. The expected utility of a 50–50 wager is negative as a result of risk aversion.

as shown in Figure 6.8. Note that $-E\{U(w)\}$ represents the payment that a person will accept *in lieu of* making this undesirable bet. The fact that $E\{U(w)\} < 0$ for a "fair" bet is referred to as *risk aversion*. This clearly requires $U(w)$ to be *concave*, as in Eq. (6.46). These simple ideas are the foundation of von Neumann and Morgenstern's (1947) axiomatic development of utility theory and can also be used as the basis for measuring utility (Hull et al. 1973). However, since we shall continue primarily to use the expected-value criterion rather than expected utility, we shall not pursue the subject further here. (Note that utility refers to discrete payments w and not to infinitesimal *flows*; this distinction is often glossed over in the literature.)

The Prior Distribution

We have seen that, in uncertain situations characterized by gambler's (or insurable) risk, the probabilities of the relevant outcomes can be assumed to be known in advance. The only uncertainty inherent in such problems is that associated with the future "roll of the dice."

In fisheries problems, as in most practical management problems, this case is exceptional. Far more prevalent is the situation in which the probabilities themselves are uncertain. Indeed, even the outcome itself may remain uncertain, owing to measurement errors, which in the case of the fishery would include annual recruitment, total catch, total effort, and the like. Model parameters and even the choice of an appropriate model may also be surrounded by uncertainty.

How are management decisions to be made in the face of such fundamental uncertainties? Three approaches that may be considered are:

1. *Deterministic or certainty-equivalent approach.* Management decisions are based upon "best estimates" of appropriate variables and parameters. Otherwise, uncertainty is ignored. The traditional MSY methodology is an example of this approach.

2. *Uncertainty-adjustment or safety-factor approach.* The deterministic approach is used to obtain a first approximation, which is then adjusted in an ad hoc fashion to allow for uncertainty. Uncertainty adjustments recommended by government fishery managers usually lie on the conservative ("safety-factor") side, an example being the $F_{0.1}$ rule (Section 3.7). The fishing industry, on the other hand, often argues for increased catch quotas *because of* uncertainty. Meetings convened to agree on annual catch quotas

may degenerate into confrontations between fishery scientists and industry representatives. The scientists are then forced to take extreme positions, even to downplay the degree of uncertainty in their estimates, in order to counter the industry's demands. A methodology devised to overcome this impasse is:

3. *Decision-theoretic approach.*† Uncertainty is explicitly recognized at the outset by means of probability distributions called (Bayesian) *prior distributions.* An objective function is identified—for example, discounted net economic yield or perhaps utility. Management decisions then attempt to maximize the mathematical expectation of this objective, relative to the prior distribution. Discussions of the decision-theoretic approach have only recently begun to appear in the fisheries literature (e.g., Walters 1975; Walters and Hilborn 1976; Mendelssohn 1980b; Ludwig and Walters 1981, 1982; Mangel and Clark 1983; Charles 1983c; Clark et al. 1984). Practical implementation awaits the future.

A (Bayesian) *prior distribution* for a parameter θ is a probability distribution $\pi(\theta)$ with the *interpretation* that

$$\mathrm{pr}(\theta_0 \leq \theta \leq \theta_0 + d\theta_0) = \pi(\theta_0)\, d\theta_0 \qquad (6.48)$$

It is usually considered that the parameter θ has an actual exact value but that this value is not known. In real applications, the actual value of θ may never become known and may be quite unknowable—for example, the current biomass B of a given fish stock or the natural mortality rate M for the same stock. Presumably B has an actual value, but M involves a level of abstraction (e.g., M may be a function of age) which implies that no "true" value of M actually exists.

The determination of an appropriate prior distribution may be difficult, particularly when there are little hard data that can be used. This problem is addressed in the decision-analysis literature (Kaufman and Thomas 1977).

In the applications to be discussed in the following section, the prior distribution is assumed to be derived from fishery data. In such cases the main problem is that of choosing the appropriate functional form for the prior distribution, after which the problem reduces to one of parameter estimation. Because of the computational complexity inherent in dynamic

---

†The term *decision analysis* is preferred by some writers, presumably to suggest that the approach is amenable to practical implementation.

decision problems, the choice of prior distribution is often dictated by the requirement of mathematical tractability. Fortunately, the influence of the assumed prior distribution diminishes over time as a result of the updating process used in dynamic decision problems.

As an example, let us consider the case where observations are taken of a random variable x having a normal distribution $N(\mu, \sigma^2)$ with known variance σ^2 but an unknown mean μ. It is shown below that the prior distribution mean μ, given n observations $x_1, x_2, \ldots, x_n$, is a normal distribution with mean $\hat{\mu} = (1/n) \sum x_n$ (the observed mean) and variance $\hat{\sigma}^2 = \sigma^2/n$. This is intuitively reasonable: The best estimate of the true mean μ is the sample mean $\hat{\mu}$, and the uncertainty associated with this estimate, as measured by $\hat{\sigma}$, decreases with n.

In other cases, the data may be so fragmentary, disorganized, or unreliable that no "scientific" approach can be used. Under such circumstances the scientist or engineer trained in classical statistical methods may simply claim that no estimate is available. Nevertheless it is almost always possible to make some "commonsense" statement about the range of possible values of the parameter in question. Suppose, for example, that the productivity q of a certain trout stream has never been assessed, but it is known that at least several hundred trout are taken each year. A maximum production per mile may be known for similar streams in the area, and experts would agree that q must fall within some definite range $q_1 \le q \le q_2$. At worst, one can then formulate the uniform prior

$$\pi(q) = \begin{cases} \dfrac{1}{q_2 - q_1} & \text{if } q_1 \le q \le q_2 \\ 0 & \text{otherwise} \end{cases}$$

The *cumulative* prior distribution $F(\theta)$ corresponding to $\pi(\theta)$ is defined in the usual way:

$$F(\theta_0) = \text{pr}(\theta \le \theta_0) = \int_{-\infty}^{\theta_0} \pi(\theta) \, d\theta \tag{6.49}$$

Clearly $F(\theta)$ is nondecreasing, with $F(-\infty) = 0$ and $F(+\infty) = 1$; see Figure 6.9.

In cases where no prior information exists, the so-called "noninformative prior" distribution

$$\pi(\theta) = 1 \quad \text{for} \quad -\infty < \theta < \infty$$

(or for $0 \le \theta < \infty$ if θ must be nonnegative) may be used, even though it is clearly not a bona fide probability distribution. Once a single updating

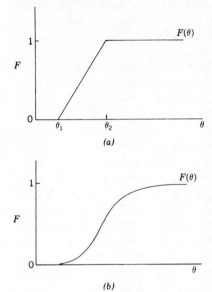

FIGURE 6.9 Cumulative prior distributions $F(\theta)$: (a) uniform prior on $\theta_1 \le \theta \le \theta_2$; (b) general case.

(see below) is performed, the noninformative prior is transformed into a proper distribution [see Eq. (6.54) below].

Optimal Decisions: Static Case

Assume now that the prior distribution $\pi(\theta)$ is specified; in general, θ will be a multidimensional parameter and π a multivariate distribution. Let A denote the set of all feasible actions or decisions, and let $V(a, \theta)$ represent the objective or payoff function corresponding to decision a and parameter θ. Thus $V(a, \theta)$ depends both on the decision $a \in A$ and on the value of the parameter θ. An *optimal decision* is defined as a decision $a^* \in A$ that maximizes the expected payoff:

$$E_\pi\{V(a^*, \theta)\} = \max_{a \in A} E_\pi\{V(a, \theta)\} \qquad (6.50)$$

This decision rule should be compared with the certainty-equivalent rule:

$$\max_{a \in A} V(a, \bar{\theta}) \qquad (6.51)$$

where $\bar{\theta}$ is the expected or estimated value of θ. Clearly, the two rules are not equivalent; in essence, the certainty-equivalent rule ignores "unlikely" outcomes in which θ differs much from $\bar{\theta}$. The certainty-equivalent rule is a poor choice whenever errors in $\bar{\theta}$ result in significant value losses.

The above formulation is static in the sense that only a single decision $a^* \in A$ is to be determined. Dynamic decision problems are characterized by two additional features. First, the objective V involves future payments, which may be affected by the history of actions taken and may also be subject to random influences. The total discounted yield from a fish stock is an obvious example. Secondly, the actions a_t taken over time may generate additional information concerning the parameter θ. In general, these future actions cannot be specified initially but depend on the actual development of the stochastic system: Action a_t is determined from the information state of the system at time t. Such decision policies are referred to as *feedback* policies.

Bayesian Updating

Suppose now that the prior distribution $\pi(\theta)$ is known. Assume that the possibility exists of conducting certain "experiments" in order to improve the estimate of θ. The outcome of such an experiment is a random variable whose distribution $f(x \mid \theta)$ depends on the true value of θ. Thus

$$f(x_0 \mid \theta_0) \, dx_0 = \text{pr}(x_0 \le x \le x_0 + dx_0, \text{ given } \theta = \theta_0) \qquad (6.52)$$

The expected value of x, given θ, is

$$\bar{x}_\theta = E\{x \mid \theta\} = \int xf(x \mid \theta) \, dx \qquad (6.53)$$

(This situation was encountered in Section 2.4, with reference to the process of searching for fish.)

We now consider the inverse question: If the experiment yields the outcome x, what does this tell us about the value of the parameter θ? Intuitively, if x differs widely from the expected outcome $\bar{x}_{\hat{\theta}}$, then we will want to adjust our estimate $\hat{\theta}$, whereas if x is close to $\bar{x}_{\hat{\theta}}$, our confidence in the estimate $\hat{\theta}$ will be increased.

Mathematically, the answer is provided by *Bayes's formula*:

$$\pi(\theta \mid x) = \frac{f(x \mid \theta) \pi(\theta)}{\displaystyle\int f(x \mid \psi) \pi(\psi) \, d\psi} \qquad (6.54)$$

Here $\pi(\theta \mid x)$ denotes the probability distribution for θ, given the observation x, and $\pi(\theta)$ is the prior distribution. The application of Eq. (6.54) is often referred to as *Bayesian updating*; the updated distribution $\pi(\theta \mid x)$ is called the *posterior* distribution. Of course, the posterior distribution may later be used as a prior distribution when subject to further updating.

An application of the updating formula (6.54) was given in Section 2.4, and a simple proof of the formula was given in the Appendix to Chapter 2. Another application, which will be used in the next section, is the following.

Let x be a random variable having a normal distribution $N(\mu, \sigma^2)$ with known variance σ^2 but unknown mean. The problem is to determine the probability distribution for μ, given that n observations $x_1, x_2, \ldots, x_n$ have been made. For $n = 0$ (no observations), the noninformative prior

$$\pi(\mu) \equiv 1, \qquad -\infty < \mu < \infty \tag{6.55}$$

is appropriate. For one observation x_1, we use the updating formula (6.54):

$$
\begin{aligned}
\pi(\mu \mid x_1) &= \frac{f(x_1 \mid \mu)\pi(\mu)}{\displaystyle\int f(x_1 \mid v)\pi(v)\,dv} \\
&= f(x_1 \mid \mu) \qquad [\text{since } \pi(\mu) \equiv 1] \\
&= n(\mu; x_1, \sigma^2)
\end{aligned}
$$

where

$$n(\mu; x, \sigma^2) = \frac{1}{\sqrt{2\pi}\,\sigma}\, e^{-(\mu-x)^2/2\sigma^2} \tag{6.56}$$

is the normal probability distribution $N(x, \sigma^2)$.

Similarly,

$$
\begin{aligned}
\pi(\mu \mid x_1, x_2) &= \frac{f(x_2 \mid \mu)\pi(\mu \mid x_1)}{\displaystyle\int f(x_2 \mid v)\pi(v \mid x_1)\,dv} \\
&= n\left(\mu; \frac{x_1 + x_2}{2}, \frac{\sigma^2}{2}\right)
\end{aligned}
$$

as follows by substituting and completion of the square. In general, it can be shown by mathematical induction that

$$\pi(\mu \mid x_1, x_2, \ldots, x_n) = n\left(\mu; \bar{x}, \frac{\sigma^2}{n}\right) \tag{6.57}$$

where $\bar{x} = (1/n)\sum x_i$ is the sample mean.

Thus (as noted earlier) after n observations, the expected value (or "best estimate") of the true mean μ is the sample mean $\bar{x}$, and the variance in this estimate is σ^2/n.

Random Sampling

As another application of Bayesian updating we consider the problem of random sampling of a population. We assume that the population is randomly distributed over an area A (or, for fish, a volume V). A subarea A_0 is sampled and found to contain n members of the population. An obvious point estimate for the total population N in A is then

$$\bar{N} = \frac{A}{A_0} n$$

but we wish to obtain a probability distribution for N.

If N is given, and $A_0/A \ll 1$, then the number of objects contained in the area A_0 is approximated by a Poisson random variable with parameter $\lambda = aN$ ($a = A_0/A$). We thus have

$$\mathrm{pr}(n \mid \lambda) = \frac{\lambda^n}{n!} e^{-\lambda} \qquad (n = 0, 1, 2, \ldots) \tag{6.58}$$

The inverse sampling problem is that of estimating λ (and hence N) from n, the number of objects observed in A_0. Bayes's formula then becomes

$$\pi(\lambda \mid n) = \frac{\mathrm{pr}(n \mid \lambda)\pi(\lambda)}{\int \mathrm{pr}(n \mid \mu)\pi(\mu)\,d\mu}$$

If we assume a noninformative prior distribution $\pi(\mu) \equiv 1$, this becomes simply

$$\pi(\lambda \mid n) = \frac{\lambda^n}{n!} e^{-\lambda} \tag{6.59}$$

This conditional distribution has its peak at $\lambda = n$, and we also have

$$E\{\lambda \mid n\} = \mathrm{var}\{\lambda \mid n\} = n + 1 \tag{6.60}$$

Thus if n objects are located in A, our estimate of N is

$$\hat{N} = \frac{n+1}{a} = \frac{A}{A_0}(n+1) \tag{6.61}$$

The coefficient of variation of this estimate is $1/\sqrt{n+1}$. [The estimate

(6.61) is biased; the unbiased estimate is n/a. However, the relative bias vanishes as n increases.]

As a simple example, suppose that .1% of an area A is sampled and yields 10 objects. The total population, as estimated by the above method, is then $\hat{N} = 11,000$, with a coefficient of variation of $1/\sqrt{11} = 30\%$. Now suppose it is desired to estimate the population with a CV of 10%. This requires $n = 101$. Hence $\hat{a} = n/\hat{N} = .0092$; that is approximately 1% of the area A should be sampled.

Stock estimates of this degree of precision are seldom if ever encountered in fisheries, not only because sampling 1%, or even .1%, of the total area of the fishing grounds is seldom feasible, but also because fish populations are never randomly (regularly) distributed. Various methods have been proposed for estimating contagiously distributed populations (Leaman 1981), the most common being stratified random sampling. In this approach, the original area A is subdivided into subareas A_i in such a way that the hypothesis of a random distribution over each A_i appears less unreasonable. Fishery data or preliminary survey data may be used in choosing the A_i.

Suppose now that posterior distributions

$$\text{pr}(\lambda_i \mid n_i) = \frac{\lambda_i^{n_i}}{n_i!} e^{-\lambda_i}$$

have been obtained for each of the areas A_i, where $\lambda_i = a_i N_i$, as before. The corresponding posterior distribution for the total population in A is then an $(n-1)$-fold convolution integral. For example, in the case of two subareas we obtain

$$\pi(N \mid n_1, n_2) = \frac{a_1^{n_1} a_2^{n_2}}{n_1! n_2!} \int_0^N t^{n_1} e^{-a_1 t} (N-t)^{n_2} e^{-a_2 t} \, dt \qquad (6.62)$$

from the formula for the distribution of the sum of two independent random variables (Heyman and Sobel 1984, p. 504).

We will not pursue the subject of sampling further here. It does seem possible, however, to address the problem of stock surveys—one of the central problems in fishery management—from the decision-theoretic viewpoint. Any attempt to assess the costs and benefits of stock surveys would probably have to be based on this approach.

Dynamic Decision Problems

Figure 6.10 shows a general schematic of an uncertain, stochastic, controlled dynamic system. The state of the system in period k is represented by X_k, and the state of information (or uncertainty) is

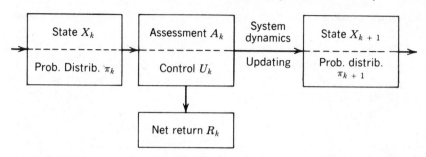

FIGURE 6.10 Schematic diagram of an uncertain, stochastic, controlled dynamic system.

represented by a multivariate probability distribution $\pi_k(\theta)$. Uncertainty may pertain to all aspects of the system, including the state X_k itself, the dynamic laws, the mean and variance of stochastic components, and so forth.

During period k assessment activities A_k (measurement, sampling) are undertaken, and controls U_k are applied, resulting in net revenues R_k (generally uncertain). The system responds to the controls through the system dynamics (uncertain, in general) to produce a subsequent state X_{k+1}. Furthermore, the assessment activities allow the probability distribution $\pi_k(\theta)$ to be updated to $\pi_{k+1}(\theta)$; in general, the amount of additional information so obtained will depend upon the amount of assessment as well as on the control and state variables.

It is assumed, ideally at least, that the system is managed (via A_k, U_k) so as to maximize some specified objective

$$ J = E\left\{ \sum_{j=1}^{N} V_j(R_j) \right\} \tag{6.63} $$

where N is the time horizon (possibly infinite), $V_j(R_j)$ denotes the present utility of future benefits R_j, and the expectation $E\{\ldots\}$ is taken over all stochastic elements involved *and* over the prior probability distribution π_1. (The updated distributions $\pi_2, \ldots$ are implicitly included, since they are determined from π_1 by the assessments, controls, and system dynamics.)

What is the nature of the problem involved in determining the optimal assessments and controls—say, in the first period? For any particular choice of A_1, U_1 the system evolves to a new state X_2, π_2 according to the uncertain, stochastic dynamics. At that stage new decisions A_2, U_2 are made, and these obviously are dependent on A_1, U_1. On the other hand, the optimal choices of A_1, U_1 clearly depend on the future choices A_2, U_2.

The way around this apparent logical circularity is the method of *dynamic programming* (or, as it is sometimes called, backward induction). First one considers the terminal period, $j = N$. For *all possible* states X_N, π_N one formulates

$$J_N(X_N, \pi_N) = \max_{A_n, U_N} E_{\pi_N}\{V_N(R_N)\} \tag{6.64}$$

Solving the indicated maximization problem determines A_N^* and U_N^* as functions of X_N and π_N. Assuming, as is generally the case, that π_N depends on known (i.e., already estimated) parameters θ_N, this maximization problem is finite-dimensional.

Proceeding to the penultimate period, we obtain

$$J_{N-1}(X_{N-1}, \pi_{N-1}) = \max_{A_{N-1}, U_{N-1}} E\{V_{N-1}(R_{N-1}) + J_N(X_N, \pi_N)\} \tag{6.65}$$

and the process can be repeated back to $J_1(X_1, \pi_1)$.

However, one should not be misled by the simplicity of the schematic notation: Except for the simplest model systems, the problem of working backward rapidly becomes computationally impossible. Suppose the Nth period problem (6.64) to have been solved by numerical computation (analytic solutions being possible only in very exceptional cases). The solution must be computed and stored for "all possible" values of X_N, θ_N—namely for a suitable discretized set of these values.

For a very coarse approximation with a simple problem, perhaps $100 \times 100 = 10^4$ such values would be required. Then at least 10^8 values would be needed at the next stage—say, several hundred megabytes of stored data. The next stage would require several *million* megabytes...

Drastic simplification is clearly needed in order to obtain a solvable problem. Only models with a small number of state variables and uncertain parameters can be treated. For such low-dimensional models, six types of simplification listed in order of decreasing restrictedness are:

1. *Deterministic method.* All random variables are replaced by their averages, and uncertain parameters by their expected values.

2. *Stochastic method.* Uncertain parameters are replaced by their expected values, but random fluctuations are retained. Expectations are taken with respect to these random variables. (This is the "insurable-risk" paradigm.)

3. *Nonadaptive decision method.* Both random and uncertain ele-

ments are retained, but future changes in uncertainty are ignored. Expectations are calculated with respect to random variables and with respect to the prior distribution.

4. *Passive adaptive decision method.* At each period, decisions are made as in the nonadaptive method. However, the existing prior distribution is always updated before the decision is actually made.

5. *Semiactive adaptive decision method.* Randomness and uncertainty are included. Calculation of the optimal policy in a given period k is based on the assumption that updating will occur only *once*, before the $(k+1)$st-period decision is made, and that a passive adaptive policy will be followed subsequently.

6. *Fully active adaptive method.* All future updating possibilities are allowed for at each stage. This is the general, computationally infeasible case. For the case of an infinite time horizon, the numerical approximation method of *value iteration* can sometimes be used [see Ludwig and Walters (1982) for a fishery application].

Note that for both the semiactive and fully active adaptive methods, the decision at any stage will be influenced by considerations of the *information* that may result from that decision; this information aspect is ignored in passive adaptive and nonadaptive methods. (The term *myopic* is sometimes used to describe decisions that ignore the value of information, but that word has already been overutilized in this book.) For example, the difference between passive adaptive and semiactive adaptive strategies was illustrated in the search model discussed in the Appendix to Chapter 2.

At this stage it is not possible to say which level of simplification is most appropriate for specific fishery applications. However, in situations where uncertainty appears to be significant, the use of an adaptive approach (levels 5 and 6) should at least be considered. Some further examples are discussed in the following section.

6.4   Fishery Applications

In this section we describe rather briefly three applications of decision theory to fishery problems. Further details can be found in the cited references.

Optimal Fishing Capacity

We first consider the case of a developing fishery in which information concerning the overall abundance and productivity of the fish stock is extremely scanty. We presume that *some* sample fishing has already taken place, demonstrating the likelihood of profitable fishery development. The question is: Should harvest capacity be controlled at the outset, and if so, what methods can be used, in view of the uncertainty involved, to estimate the optimal capacity initially?

Perhaps the most common development policy in practice is based on laissez-faire. The fishery is allowed to develop under private, competitive profit motives. Regulation is only contemplated when the fishing industry begins to show signs of stress. By then, serious problems of overcapacity may have developed, rendering the necessary catch restrictions all the more difficult to achieve (Gulland 1978). Let us recall, however, that the optimal management policy for an overcapitalized fishery is generally different than that for an undercapitalized fishery—see Section 3.3.

Fishery development is frequently actively encouraged by means of government subsidies for vessel and plant construction. As discussed in Section 6.2, such subsidies may be warranted in order to counter the risk aversion of private investors. However, it is not only risks, but capital costs themselves, which are reduced by capital subsidies for vessel and plant construction. Unless such programs are combined with rigorous control of the expansion of fishing capacity, severe overexpansion is virtually ensured. The importance of attempting to determine optimal capacity during the early stages of development thus becomes obvious.

Consider first the following highly simplified model, designed to analyze the effects of the uncertainty which is involved in estimating the average sustainable yield (Clark et al. 1985). The model also provides a pedagogical illustration of the various simplified decision techniques listed in the previous section.

Recruitment levels R_k in year k are assumed to be independent random variables with common distribution $f(R)$, which is independent of k and of spawning escapement from previous years. This assumption will be relaxed later.

Specifically, we assume that R_k is lognormally distributed, so that $X_k = \log R_k$ is normally distributed with mean μ and variance σ^2. It follows that

$$\bar{R} = e^{\mu + \sigma^2/2}, \qquad \sigma_R^2 = \bar{R}^2(e^{\sigma^2} - 1) \qquad (6.66)$$

or inversely,

$$\sigma^2 = \log\left(1 + \frac{\sigma_R^2}{\bar{R}^2}\right), \qquad \mu = \log \bar{R} - \frac{\sigma^2}{2}$$

The variance σ^2 of X is thus determined by the coefficient of variation $\sigma_R / \bar{R}$ of recruitment R. We make the further simplification of assuming that this variance is known. The mean μ, however, will be treated as an uncertain parameter.

Suppose that recruitment has been observed for a period of m years. Let $X_1, \ldots, X_m$ denote these observations transformed logarithmically; $X_i = \log R_i$. Then, as shown in the preceding section, Eq. (6.57), an appropriate prior distribution for the mean μ is

$$\pi_0(\mu) = \pi_0(\mu \mid X_1, \ldots, X_n) = n(\mu; \mu_0, \sigma_0) \qquad (6.67)$$

where $\mu_0 = \bar{X} = \sum_i^m X_i / m$ is the sample mean (and the prior estimate of μ), and $\sigma_0^2 = \sigma^2 / m$. This assumes that the observations X_i are exact; if in fact the X_i involve observation errors that are normally distributed with mean 0 and variance σ_ϵ^2, then the prior distribution (6.67) should be replaced by $n[\mu; \bar{X}, (\sigma^2 + \sigma_\epsilon^2)/m]$.

Before proceeding, let us discuss briefly the assumption of lognormal recruitment. First, the lognormal distribution has been shown to provide an acceptable fit to recruitment data from a large number of diverse fisheries (Hennemuth et al. 1980, Beddington and Cooke 1983). From the theoretical point of view, an approximately normal distribution for $X_k = \log R_k$ would be implied by the central limit theorem of probability theory, provided recruitment R_k is expressible as a *product* of numerous independent random variables. Inasmuch as the recruiting population may have been subject to a series of independent mortality events over time, this latter hypothesis does have some biological foundation. Of course, besides having both empirical and theoretical support, the lognormal distribution is convenient analytically.

Our fishery model will also ignore age structure, so that the annual recruitment R_k in fact represents the total stock (biomass) available to the fishery in year k [for an extension of the model including age structure, see Clark et al. (1985)].

Next we adopt two further simplifying assumptions: First, fishing mortality is assumed to be directly proportional to fishing effort (type II concentration), and second, fishing effort is directly proportional to fishing capacity. Thus existing capacity is always fully utilized—a not unreasonable assumption for a developing fishery (see Section 6.1 for a more general model of this aspect). If F denotes fishing capacity, measured in terms of instantaneous fishing mortality, then the relationship between recruitment and annual catch becomes simply

$$C_k = R_k(1 - e^{-F}) \tag{6.68}$$

where natural mortality during the fishing season is neglected.

Consider first the nonadaptive control problem. We wish to determine the initial capacity level F_0 that will maximize expected net returns, allowing for uncertainty in the parameter μ but not allowing for future updating. This problem can be expressed as follows:

$$\underset{F_0 \geq 0}{\text{Maximize}} \left[E\left\{ \sum_{k=1}^{H} \alpha^k p R_k (1 - e^{-F_0}) \right\} - cF_0 \right] \tag{6.69}$$

where the symbols H, α, p, and c have obvious significance; these parameters are assumed to be known constants. Depreciation is here ignored; since capacity is assumed to be fully utilized, variable operating costs are proportional to F_0 and are thus included in the cost coefficient c, which also includes fixed capacity costs.

Because of the independence of R_k, Eq. (6.69) can be rewritten as

$$\underset{F_0 \geq 0}{\text{Maximize}} [A(H)(1 - e^{-F_0}) E\{R_k\} - cF_0] \tag{6.70}$$

where

$$A(H) = p \sum_{k=1}^{H} \alpha^k \tag{6.71}$$

The expectation $E\{R_k\}$ can be calculated explicitly:

$$E\{R_k\} = E_{\pi_0}\{E\{R_k \mid \mu\}\}$$

where π_0 is given by Eq. (6.67), or

$$E\{R_k\} = E_{\pi_0}\{e^{\mu + \sigma^2/2}\} \qquad \text{[by Eq. (6.66)]}$$

$$= \int_{-\infty}^{\infty} e^{\mu + \sigma^2/2} n(\mu; \mu_0, \sigma_0^2) \, d\mu$$

$$= e^{\mu_0 + (\sigma^2 + \sigma_0^2)/2} \tag{6.72}$$

When this expression is substituted into Eq. (6.70), we obtain a trivial maximization problem with the solution

$$F_0^* = \log \frac{A(H)}{c} + \mu_0 + \frac{\sigma^2 + \sigma_0^2}{2} \tag{6.73}$$

if the expression on the right is positive; otherwise $F_0^* = 0$. Clearly, the latter possibility will arise if the cost/price ratio c/p is sufficiently large.

It is instructive to compare this solution with the certainty-equivalent solution, obtained by replacing the uncertain parameter μ by its prior estimate μ_0 at the outset. (Because of the linearity of the present model, the certainty-equivalent solution is identical with the deterministic solution.) For this case we have

$$E\{R_k\} = E\{R_k \mid \mu_0\} = e^{\mu_0 + \sigma^2/2}$$

and the corresponding optimal capacity is therefore

$$F'_0 = \log \frac{A(H)}{c} + \mu_0 + \frac{\sigma^2}{2} \tag{6.74}$$

We see that the passive adaptive method for handling recruitment uncertainty leads to a *larger* estimate of optimal capacity than does the certainty-equivalent method:

$$F_0^* = F'_0 + \frac{\sigma^2}{2m} \tag{6.75}$$

As a numerical illustration, Clark et al. (1985) consider the following values adapted from the tropical prawn model of Clark and Kirkwood (1979):

$$R_1 = 1.5 \times 10^7 \text{ kg}$$

$$R_2 = 0.32 \times 10^7 \text{ kg}$$

$$R_3 = 0.55 \times 10^7 \text{ kg}$$

$$\sigma = .58$$

$$\frac{c}{p} = 2.2 \times 10^7 \text{ kg}$$

$$\alpha = .91$$

$$H = 20$$

Using the above three recruitment values, we have

$$\mu_0 = 15.67$$

and we find that

$$F'_0 = 1.08, \qquad F_0^* = 1.13$$

These values are fairly close to one another, since the effect of uncertainty, $\sigma^2/6$, appearing in Eq. (6.75) is fairly small relative to the optimal F_0 value itself.

It may seem counterintuitive that the consideration of uncertainty should necessarily lead to an increase in fishing capacity. This is the result of various features of our model, including the lack of any stock-recruitment effect as well as the exclusion of risk aversion. But it is also characteristic of the passive adaptive method and not of the semiactive adaptive method, as we now show.

Assume that an initial capacity F_0 will be selected, but that after a certain number N years of experience the original recruitment estimate will be updated and additional capacity F_1 will be obtained if indicated by the revised estimate. The process may in actuality be repeated, but the semiactive technique ignores any further updating.

We continue to assume that capacity once obtained is fully utilized thereafter; in particular, capacity is nonmalleable. Intuitively, one would expect that uncertainty about long-term recruitment prospects would call for a conservative initial capacity, since an adjustment *upward* can be made later. This turns out to be correct.

Let $J_1(F_0, \mu')$ denote the maximum expected return, over the residual horizon $H - N$, given that initial capacity F_0 exists and given the updated estimate μ'. From the above calculation we obtain

$$J_1(F_0, \mu') = A(H - N)e^{\mu' + (\sigma^2 + \sigma'^2)/2}(1 - e^{-(F_0 + F_1^*)}) - cF_1^* \qquad (6.76)$$

where

$$F_1^* = \log \frac{A(H - N)}{c} + \mu' + \frac{\sigma^2 + \sigma'^2}{2} - F_0 \qquad (6.77)$$

if this expression is positive, and $F_1^* = 0$ otherwise. Here

$$\sigma'^2 = \frac{\sigma^2}{m + N} \qquad (6.78)$$

is the variance of the posterior distribution.

The dynamic programming equation [cf. Eq. (6.65)] now becomes

$$J_2 = \max_{F_0 \geq 0}[A(N)e^{\mu_0 + (\sigma^2 + \sigma_0^2)/2}(1 - e^{-F_0}) + \alpha^N E_{\pi_0} E_\mu\{J_1(F_0, \mu')\}] \qquad (6.79)$$

It turns out that the double expectation here can be simplified and expressed in terms of standard Fortran functions (namely, the error function), and hence this expectation can be computed efficiently. Thus the optimal initial capacity level for the semiactive case, say F_0^{**}, is easily determined [see Clark et al. (1985) for details]. For the above numerical values we obtain (for $N = 3$, which is the best choice for N)

$$F_0^{**} = 1.05$$

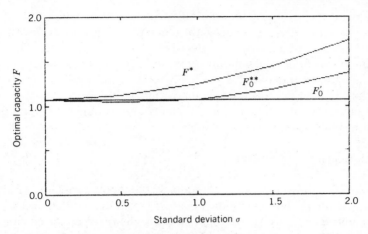

FIGURE 6.11   Optimal initial capacity levels F'_0, F_0^{**}, and F_0^* as functions of the standard deviation σ (average recruitment R = constant). (From Clark et al. 1984.)

which should be compared with the previous values

$$F'_0 = 1.08, \qquad F_0^* = 1.13$$

This confirms our intuitive feeling that allowing for updating should reduce F_0. Indeed, for these parameter values we obtain $F_0^{**} < F'_0$; that is, inclusion of uncertainty via the semiactive adaptive method *reduces F_0* relative to the certainty-equivalent result rather than increases it.

The relationship between the different F_0 values depends on the degree of uncertainty about μ, as represented by the variance σ^2. This is illustrated in Figure 6.11; here the long-run average recruitment $R = e^{\mu + \sigma^2/2}$ is kept constant (by varying μ along with σ), so that the deterministic solution F'_0 also remains constant.

Note that, for moderate levels of recruitment variability (say, with coefficient of variation $\leq 100\%$) the initial capacity decision for this model is not highly sensitive either to σ or to the decision paradigm that is employed. Furthermore, small relative errors in capacity result in even smaller relative losses in the expected economic return.

This suggests that straightforward deterministic models may be appropriate after all for capacity decisions. Even a small series of recruitment data (here $m = 3$ values) seems sufficient to formulate a fairly good decision. The data themselves need not be highly accurate. The assumption that no stock-recruitment relationship exists, while probably reasonable for the case of tropical prawn fisheries, should obviously be questioned in other examples (see below).

Perhaps the main conclusion of this analysis is that quite good capacity decisions can often be made with a minimum of data and with quite simple models by employing a decision-theoretic approach. The traditional approach of avoiding any decision on the grounds that "available data are too fragmentary" is not likely to be justifiable—especially since this laissez-faire approach leads almost inevitably to overcapacity. (The fact that recruitment is *sometimes* high enough to yield short-term profits is no justification for excess capacity.)

Stock-Recruitment Effects

One of the most fundamental and most difficult problems of fishery management is the estimation of stock-recruitment relationships. The form and uncertainty of this relationship is clearly of major economic as well as biological importance. Yet fishery economics has almost universally ignored the problem, just as fishery biology has paid little attention to the economic implications. Because of the dominance of uncertainty, a decision-analytic approach seems essential.

Among the features that lead to severe statistical difficulties in stock-recruitment estimation are:

1. Paucity of data

2. Inaccuracy of data

3. Unreliability of data

4. Lack of variation in stock levels

5. Extreme variation in recruitment levels

6. Uncertainty as to the "correct" model

An example of stock-recruitment data (for southern bluefin tuna) is shown in Figure 6.12 (see also Figures 2.9 and 3.16). Note that the data plotted in this diagram has been derived from catch data by means of cohort analysis, so that the data themselves are subject to errors from this process (this is not shown in the diagram).

The data in Figure 6.12 display limited variation in spawning stock size, and it is clear that this feature is likely to engender high uncertainty in any fitted stock-recruitment model. As noted by Ludwig (1982), this lack of variation is a common feature of fishery data and may be intensified by management practices which aim at constant escapement levels. We discuss this issue further below.

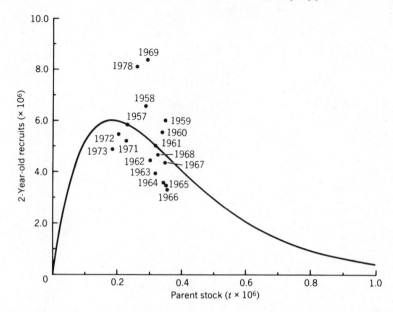

FIGURE 6.12   Ricker curve fitted to stock and recruitment data for southern bluefin tuna (Murphy 1982). Stock and recruitment estimates were derived by means of cohort analysis.

The data shown in Figure 2.9 show much greater variation in stock levels (but the spawning stock data are strongly serially correlated). Nevertheless, considerable uncertainty obviously remains. However, a strong correlation between stock and recruitment obviously exists *at low stock levels*. The latter is a common feature of stock-recruitment data for many heavily exploited fish stocks. In many cases, the approximately linear section of the curve appears to cover roughly 20% of the range of historical spawning stock levels (Beddington and Cooke 1983).

For obvious reasons, there is little hope of estimating a stock-recruitment relation for an undeveloped or slightly developed fishery. However, the simplification of assuming that no such relationship exists is clearly extreme. The only way around this impasse seems to be the adoption of some ad hoc recruitment model.

For example, Clark et al. (1985) adopt a "linear threshold" model of the form

$$R_{k+1} = Z_k \phi(S_k) \tag{6.80}$$

where Z_k are independent, identically distributed random variables and

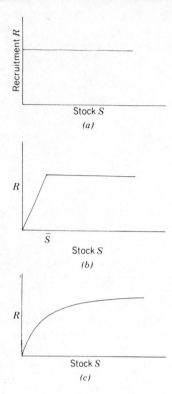

FIGURE 6.13 Stock-recruitment models: (a) constant (expected) recruitment; (b) linear threshold model; (c) Cushing model.

$\phi(S)$ is the threshold function

$$\phi(S) = \min\left(\frac{S}{\bar{S}}, 1\right) \qquad (6.81)$$

(see Fig. 6.13b). The threshold $\bar{S}$ is specified by Clark et al. (1985) as 20% of the estimated long-term average spawning biomass level.

Given that $\phi(S)$ has been specified in this ad hoc fashion, the complete stochastic stock-recruitment relation (6.80) is determined once the probability distribution of Z is specified. Clark et al. (1985) assume that Z is lognormally distributed, with uncertain mean μ and known variance σ^2. Thus the prior and posterior distributions for μ are the same as above (at least assuming $S_k \geq \bar{S}$; recruitment levels below $\bar{S}$, if any, can simply be deleted from the data set used to estimated μ). Computations of expected return are then carried out by Monte Carlo simulation; the reader is referred to the original paper for further details.

An alternative, analytic stock-recruitment model, due to Cushing

(1971), is the following:

$$R_{k+1} = S_k^\theta \cdot Z_k \qquad (6.82)$$

where S_k denotes spawning escapement and Z_k is a random variable, which we again assume lognormal, $\log Z_k \sim N(\mu, \sigma^2)$ with uncertain mean μ but known variance σ^2. The exponent θ, which describes the strength of the stock-recruitment relationship, is also assumed to be known; this is not to be thought of as a realistic assumption, but as a method of obtaining a sensitivity analysis of our first, simplified model which corresponds to the case $\theta = 0$. From Eq. (6.82) we have

$$E\{R_{k+1}\} = S_k^\theta e^{\mu + \sigma^2/2} \qquad (6.83)$$

Introducing the notation

$$x_k = \log R_k, \qquad y_k = \log Z_k$$

we obtain the equation

$$x_{k+1} = \theta \log S_k + y_k \qquad (6.84)$$

with y_k being normal $N(\mu, \sigma^2)$. Suppose we possess historical data on stock and recruitment, namely, $x_0, x_{-1}, \ldots, x_{-N_0+1}, S_{-1}, \ldots, S_{-N_0}$. Write

$$q_{-i} = x_{-i+1} - \theta \log S_{-i}$$

Then the prior distribution for μ is $N(\hat{\mu}_0, \sigma_0^2)$, where

$$\hat{\mu}_0 = \frac{1}{N_0} \sum_1^{N_0} q_{-i} \quad \text{and} \quad \sigma_0^2 = \frac{\sigma^2}{N_0} \qquad (6.85)$$

The updating formulas are the same as before, with q_i replacing $\log R_i$.

· The optimal capacity problems, for nonadaptive and semiactive adaptive controls, are now formulated exactly as before. Because of the presence of the stock-recruitment effect, however, an analytic solution is no longer possible.

Clark et al. (1985) performed Monte Carlo simulations and numerical integration (requiring a considerable amount of computer time) in order to determine the optimal F_0 values. Using an ad hoc value $\theta = 0.22$ and the same parameter values as before, they obtained the results

$$F_0^* \cong 0.70, \qquad F_0^{**} \cong 0.60$$

These values should be compared with the values determined from the same data, but assuming no stock-recruitment effect:

$$F_0^* = 1.13, \qquad F_0^{**} = 1.08 \quad 1.05 \qquad \rightarrow p.\ 252$$

In both cases, the nonadaptive method (no updating allowed for) calls for a larger capacity than does the semiactive adaptive method. The stock-recruitment assumption, however, severely reduces the optimal capacity levels in both cases. This is not surprising, inasmuch as the assumed stock-recruitment relationship has the effect of reducing the average recruitment as the escapement declines. Consequently the fishery is less profitable at any level of capacity, implying a reduction in the optimal capacity.

Figure 6.13 shows the deterministic stock-recruitment curves corresponding to the three models described here. Which model is most appropriate for the case of an undeveloped fishery? This choice would have to be made by the scientists involved on the basis of which form has proven successful for other stocks of similar species.

The threshold and Cushing models each involve a single parameter θ (or $\bar{S}$). A decision-theoretic approach would treat θ as an uncertain parameter (Charles 1983c), and an appropriate prior distribution $\pi(\theta)$ would be adopted using standard procedures of decision analysis.

Eventually the statistical estimation of a stock-recruitment relation from fishery data might become feasible. We next present a brief discussion of this problem from the decision-theoretic viewpoint.

Adaptive Fishery Management

The material in this section is adapted from the work of C. J. Walters, D. A. Ludwig, and R. Hilborn, colleagues of mine at the University of British Columbia. Because of their special interest in Pacific salmon, most of the published work pertains to the Ricker stock-recruitment model, which was developed specifically for salmon. With some exceptions (Chuma 1981), the Pacific salmon species are characterized by nonoverlapping generations, so that cohort structure can be ignored in the models.

The Ricker model in stochastic form (Ludwig and Walters 1982) is

$$R_{k+1} = S_k e^{\alpha + \beta S_k + \sigma w_k} \tag{6.86}$$

where S_k denotes spawning escapement, R_{k+1} subsequent recruitment, and the random variables w_k are assumed normal $N(0, 1)$. The Ricker stock-recruitment function

$$G(S) = S e^{\alpha + \beta S} \tag{6.87}$$

exhibits overcompensation—see Figure 6.14. The natural equilibrium is at $\bar{S} = -\alpha/\beta$. The peak of the curve is at $S = -1/\beta$, which lies to the left of $\bar{S}$ if and only if $\alpha > 1$ [Fig. 6.14(a)]. The slope at the origin is $G'(0) = e^{\alpha}$, which represents the recruit-to-spawner ratio for low escape-

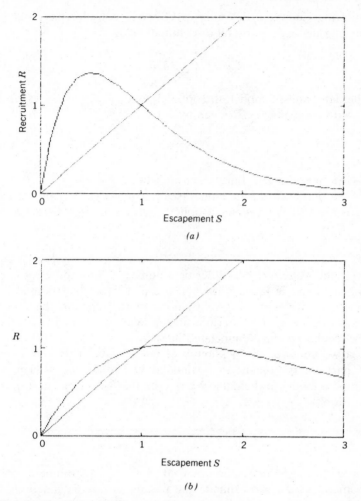

FIGURE 6.14 The Ricker stock-recruitment curves: (a) $\alpha > 1$, (b) $\alpha < 1$ (equilibrium escapement normalized at $\bar{S} = 1$).

ment levels. A highly fecund species will therefore have $\alpha \gg 1$ and (if the Ricker model is "correct") will exhibit significant overcompensation.

Note that taking expectations in Eq. (6.86) does not yield (6.87), but rather

$$\bar{R}_{k+1} = Se^{\alpha + \sigma^2/2 + \beta S} \tag{6.88}$$

Thus the "certainty equivalent" of the stochastic model should have α replaced by $\alpha + \sigma^2/2$.

Following Ludwig and Walters (1982), we consider the objective of maximizing the expected total discounted yield:

$$V(R_0) = E\left\{\sum_{k=0}^{\infty} \delta^k(R_k - S_k)\right\} \tag{6.89}$$

For the deterministic model, the optimal policy is a constant-escapement policy with escapement S_D^* given by

$$G'(S_D^*) = \frac{1}{\delta} \tag{6.90}$$

where the subscript D means "deterministic."

For the purely stochastic model (see Section 6.1), the optimal harvest policy is a constant-escapement policy, with target escapement S_0^* satisfying

$$S_0^* \geq S_D^*$$

Ludwig and Walters (1982) describe a numerical procedure for computing S_0^* (see also Clark and Kirkwood 1984). These calculations show that in fact, even for quite large variations in recruitment, the difference between stochastic and deterministic optimal escapements is minor, in support of the principle enunciated in Section 6.1.

Next we consider the possibility of uncertainty in the stock-recruitment parameters, which are estimated by fitting the Ricker stock-recruitment model to the data. We rewrite the Ricker model, Eq. (6.86), in the form

$$Y_k = \log \frac{R_{k+1}}{S_k} = \alpha + \beta S_k + \sigma w_k \tag{6.91}$$

Given annual (or generation-period) data for recruitment R_k and escapement S_k, we can estimate the parameters α, β, and σ by linear regression. Arguing that the center of mass $(\bar{S}, \bar{Y})$ of the data points is usually well determined, relative to the slope β of the regression line, Ludwig and Walters replace Eq. (6.91) by

$$R_{k+1} = S_k \exp[\bar{Y} + \beta(S_k - \bar{S}) + \sigma w_k] \tag{6.92}$$

in which the only uncertain parameter is considered to be β. They assume a normal prior distribution $\pi_0 \sim N(\hat{\beta}, \hat{\sigma}_0^2)$ for β and consider three types of decision rules: nonadaptive, passive adaptive, and active adaptive (Ludwig and Walters employ different terminology).

The nonadaptive rule (NA) simply involves averaging in addition over the prior distribution for β:

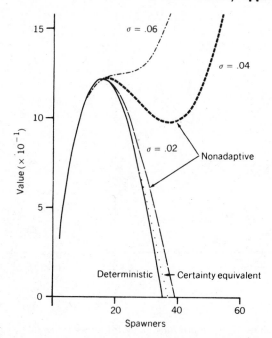

FIGURE 6.15 Values of the return function V depending on spawning escapement, for deterministic (D), certainty-equivalent (CE), and nonadaptive (NA) decision rules. Here $\hat{\beta} = 0.3$, and $\hat{\sigma} = .02$, .04, and .06. (From Ludwig and Walters 1982.)

$$V_{NA}(R_0) = E_{\pi_0} E_w \left\{ \sum_0^\infty \delta^k (R_k - S_k) \right\} \qquad (6.93)$$

Not surprisingly, this extra averaging has very little effect; we obtain $S_{NA}^* \approx S_D^*$ unless $\hat{\sigma}_0 > |\hat{\beta}|$. Indeed, if a constant-escapement policy with escapement S is used, the resulting value $V(R_0; S)$ is only slightly altered in passing from deterministic to certainty-equivalent (CE) to NA formulations, unless the uncertainty in the parameter is large—see Figure 6.15. (For $\hat{\sigma}_0 > |\hat{\beta}|$ the probability that $\beta > 0$ becomes significant, in which case the model predicts *infinitely* large values V as $S \to +\infty$.)

The value functions V for passive adaptive and (fully) active adaptive strategies are shown in Figure 6.16. Here S represents spawning escapement in the first year only; future updating of the parameter β may lead to alterations in subsequent escapement policy.

We recall that the passive adaptive method assumes that the escapement decision S^* is made in each period on the basis of the most recently updated parameter estimates. Under the active adaptive method, on the

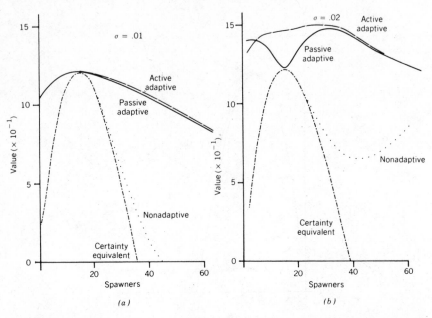

FIGURE 6.16 Values of the return function V for nonadaptive (NA), passive adaptive (PA), and active adaptive (AA) strategies: (a) $\hat{\sigma} = .01$; (b) $\hat{\sigma} = .02$ (From Ludwig and Walters 1982.)

other hand, the current escapement decision also takes into account the possibility of improving future parameter estimates (i.e., reducing uncertainty).

Under fairly low initial parameter uncertainty ($\hat{\sigma}_0 = .01$), the passive and active adaptive strategies give similar values, and the optimum first-year escapement is still approximately at the nonadaptive level S_{NA}^* (Fig. 6.16a). For escapement levels $S^* \neq S_{NA}$, however, the *value* of the adaptive strategies is significantly higher than that of the nonadaptive strategies. This difference in value can be understood as the value of *information* resulting from using an "experimental" escapement level, differing widely from the apparently optimal escapement S_{NA}^*. The more that S differs from S_{NA}^* the greater is the degree of improvement in the accuracy of the parameter estimate.

When the level of uncertainty is increased (Fig. 6.16b), the anticipated value of information obtained from such an "experimental" escapement dominates the expected value from exploiting the resource. Note that the expected payoff for the passive adaptive method is *minimized* at $S = S_{NA}^*$, which reflects the fact that little or no new information is derived from this escapement policy. The active adaptive method does not produce a minimum at S_{NA}^* because the computation assumes that appropriate action in varying S will be taken in year 2 and subsequent years.

The reader is referred to the original article (Ludwig and Walters 1982) for a more complete discussion, particularly of the numerical algorithms employed in the computation.

These results lead to the following tentative conclusions. First, at "low" levels of parameter uncertainty, simple deterministic models may lead to management decisions that are close to the optimum. But for "moderate" or "high" levels of uncertainty the deterministic approach may be quite inadequate. In order to reduce the level of uncertainty it may be necessary temporarily to employ management policies that diverge widely from the deterministic estimate.

What constitutes "low" vs. "high" uncertainty is necessarily vague. Our experience with various models suggests that uncertainty can be considered "low" whenever coefficients of variation are below 50%, and "high" if they exceed 100%. Whether such a characterization holds up under further investigation remains to be seen.

The Value of Stock Assessments

In every model discussed so far in this book it has tacitly been assumed that no uncertainty pertains to the actual stock level X_k itself. Thus, harvest policies that specify a particular escapement S_k, constant or otherwise, presuppose that escapement can in fact be accurately monitored and controlled. But this is extremely unlikely—the accurate assessment of marine fish stocks is one of the more difficult and expensive aspects of fishery management. Even with carefully designed, thorough annual stock assessments, it is uncommon to obtain estimates of stock abundance with narrow confidence limits.

Such uncertainty often results in controversy, as fishery biologists and industry representatives take opposing views regarding the proper impact of stock uncertainty on allowable catches. The scientist feels obliged to recommend ever-increasing amounts of government funds to be devoted to improving the accuracy of stock assessments in order to justify the recommended catch limitations.

The decision-theoretic approach offers a possible resolution to this impasse, which has many features in common with other resource or environmental controversies to which these methods have been applied. Many details remain to be worked out before implementation will become practical. Here we follow an initial approach due to Clark and Kirkwood (1984).

Once again we employ the stochastic stock-recruitment model

$$X_{k+1} = Z_k G(S_k) \qquad (6.94)$$

$$S_k = X_k - H_k \qquad (6.95)$$

with the objective of maximizing the discounted value of the harvests:

$$\text{Maximize } E\left\{ \sum_{k=1}^{\infty} \alpha^{k-1} H_k \right\} \qquad (6.96)$$

We return to the notation of Section 6.1, assuming that the random variables Z_k are independent and identically distributed, with distribution $f(z)$ and mean $\bar{Z} = 1$.

In order to concentrate on the effects of uncertainty in the stock levels X_k, we will suppress other sources of uncertainty and assume that both the stock-recruitment function $G(S)$ and the distribution $f(z)$ are known. In addition, we will suppose that annual escapement S_k can be accurately determined after the close of each fishing season. This is a reasonable approximation to the truth for certain stocks of Pacific salmon for which a count of fish on spawning streams can be obtained. However, the assumption of known escapement is an oversimplification for most other species, although data obtained from the fishery itself should in principle yield quite good estimates of escapement.

First we consider the case in which recruitment X_k cannot be monitored at all but quotas Q_k must nevertheless be determined at the outset of the fishing season. *Forecasts* of recruitment can be made on the basis of the previous escapement. Namely, since Z_k has known distribution $f(z)$, and since $g_k = G(S_k)$ is known, recruitment $X_{k+1} = Z_k g_k$ has prior probability distribution

$$\pi(x) = \frac{1}{g_k} f\left(\frac{x}{g_k}\right) \qquad (6.97)$$

Let $J_n(S_0)$ denote the maximum expected return, as in Eq. (6.95), but with a time horizon of n periods and with escapement S_0 from the preceding season. If Q_1 denotes the quota for the first period, we assume that the actual catch H_1 equals Q_1, unless $X_1 < Q_1$, in which case $H_1 = X_1$. Thus

$$H_1 = \langle X_1, Q_1 \rangle = \min(X_1, Q_1) \qquad (6.98)$$

This somewhat extreme formulation emphasizes the danger of misspecifying the quota when stock levels are uncertain.

The value functions $J_n(S_0)$ satisfy the recursion (dynamic programming) equation

$$J_{n+1}(S_0) = \max_{Q \geq 0} E_{\pi_1}\{\langle X_1, Q \rangle + \alpha J_n(G(X_1 - \langle X_1, Q \rangle))\} \qquad (6.99)$$

This follows easily, upon noting that

$$S_1 = X_1 - H_1 = X_1 - \langle X_1, Q \rangle$$

This equation may be compared with the case of Eq. (6.8), in which recruitment X_1 is assumed to be known at the outset of the season:

$$\bar{J}_{n+1}(X_1) = \max_{0 \leq H_1 \leq X_1} [H_1 + \alpha E_Z\{\bar{J}_n(ZG(X_1 - H_1))\}] \qquad (6.100)$$

We recall that, by making the substitution $S_1 = X_1 - H_1$ in Eq. (6.100), we were led immediately to the conclusion that the optimal harvest policy is a constant-escapement policy. This substitution does not apply to Eq. (6.99), and the optimal policy is no longer a constant-escapement policy. Expected escapement $S_1 = E_\pi\{X_1 - \langle X_1, Q^* \rangle\}$ now depends upon the previous escapement S_0—see Figure 6.17.

Given that recruitment X_1 is unknown and that the resource stock will be fished to extinction if in fact $Q > X_1$, it would appear likely that an optimal quota policy would be highly conservative compared, say, to the certainty-equivalent deterministic optimum. Numerical solution of Eq. (6.99) confirms that this is indeed the case if the variance σ^2 of log recruitment is not too large (Clark and Kirkwood 1984). However, if σ^2 is large (coefficient of variation greater than about 80%), then the optimal quota may result in extinction with high probability if the expected recruitment $\bar{X}_1 = G(S_0)$ is large. This phenomenon persists even for zero discounting and an arbitrarily long time horizon. The explanation is this: If $\bar{X}_1$ and σ^2 are both large, then only a small quota Q will provide a high probability of stock protection (since X_1 may in fact

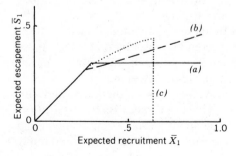

FIGURE 6.17 Optimal expected escapement $\bar{S}_1$ vs. expected recruitment $\bar{X}_1$ under recruitment uncertainty: (a) coefficient of variation $= 0$; (b) coefficient of variation $= 50\%$; (c) coefficient of variation $= 100\%$. (From Clark and Kirkwood 1984.)

be small), but a small quota inhibits large catches if in fact X_1 is large. In order to maximize expected total long-run catch, one must take some chance of eliminating the stock entirely.

This result suggests, on the one hand, that explicit stock-protection objectives and policies may be necessary adjuncts to economic goals—even with zero discounting! But it also emphasizes the importance of "on-line" stock assessments and close monitoring of highly fluctuating stocks, which may be fished nearly to extinction before the crucial evidence eventually reaches the desk of the fishery manager.

The above framework can be used to obtain a preliminary estimate of the value of such on-line stock assessments. The ideal assessment would result in completely accurate assessments of recruitment X_1 prior to the quota decision Q_1. This is in fact the assumption underlying the model of Eq. (6.100). The expected increase in net return resulting from such ideal assessments every year can therefore be written as

$$V_n(S_0) = E_\pi\{\tilde{J}_n(ZG(S_0))\} - J_n(S_0) \tag{6.101}$$

The numerical calculations of Clark and Kirkwood indicated that the increase in expected value resulting from stock assessment may approach 100% of the nonassessment value (if, for example, the coefficient of variation of recruitment is 100%). Since in fact stock surveys also yield information about escapement levels as well as stock-recruitment relationships (and other biological information), their real value is doubtlessly even greater than indicated by our simplified analysis. In view of both the importance and the expense of stock assessments, professional evaluation of their expected costs and benefits seems overdue.

Environmental Influences

The factors Z_k in the stochastic stock-recruitment model, Eq. (6.94), have been treated as purely random and unpredictable "noise" factors affecting annual recruitment. It is widely believed, however, that recruitment is strongly influenced by environmental factors of various kinds, including sea temperature, currents, river runoff, and so on. The values of Z_k are determined in part by these environmental factors. If so, it follows that recruitment predictions could be improved provided these environmental relationships were understood and that the relevant environmental variables themselves were regularly monitored.

It is often asserted that such improvements in prediction would also greatly improve fishery management (Ocean Sciences Board 1980, Frye 1983), but how this might actually come about has not been delineated in any detail (Walters 1984). Improved recruitment predictions would imply a reduction in the variance of the prior distribution $\pi(x)$ of Eq. (6.97). More precisely, the original prior distribution $\pi(x)$ would be updated to take account of the environmental information. Thus the environmental information would have the same effect as stock assessment in terms of reducing uncertainty regarding current recruitment levels. Indeed, the two updating procedures could be used in conjunction, the latter obviously being more direct in terms of the actual fish stock. From an economic viewpoint, it would be possible in principle to determine the cost-effectiveness of the two approaches, as noted above for the case of stock assessments.

The use of environmental data might also result in an increase in the lead time of forecasts. This could be particularly valuable in potentially catastrophic situations, such as El Niño events, by providing management authorities with advance warning of greatly reduced stock levels. The value of predictions of unusually high recruitment is less obvious, especially if such predictions turn out to be rather unreliable. Temporarily high stock levels can lead to the usual problem of market gluts, on the one hand, or can induce rapid expansion of fishing capacity in the case that markets are receptive. Advance notice would do little to relieve the first of these problems, but might provide the management agency with an incentive to be especially vigilant in controlling new entry.

To summarize, while oceanographic environmental data might allow for improved short-term predictions of recruitment, the value of this information in fishery management may be somewhat limited. Improved long-term predictions would require correspondingly long-term environmental predictions, which are well beyond present-day technology.

6.5 Summary and Perspective

The treatment of fluctuations and uncertainty in fisheries systems given in this chapter is far from comprehensive. Most of the models discussed here are highly simplified, and the results must be interpreted with these simplifications in mind.

The distinction between the modelling of fluctuations and of uncertainty has been emphasized. Fluctuations induce uncertainty concerning both the future "states of nature" that will be encountered and the underlying properties of the system under investigation, which must be inferred from data samples. Information about the system is itself a dynamic variable, which is usually subject to some degree of control. Information has an economic value, in the sense that expected benefits from the resource may be increased by means of the deft collection and handling of data and information.

These considerations lead to the conclusion that decision-theoretic methods may be needed in fishery management. Unfortunately, most fishery managers, having been trained as biologists, have no background in decision analysis. The classical statistics taught in today's biology curricula were designed for basic scientific studies (hypothesis testing) and are largely inappropriate for management applications. This assertion is supported by even the most cursory perusal of the current fisheries literature, in which the need for improved scientific understanding is universally stressed but with little or no suggestion of how such understanding would actually aid fishery management.

A decision-theoretic approach to fishery management also has the potential for resolving the perennial dispute between fishery managers and the fishing industry. The current "scientific" approach to fishery statistics simply leads to an impasse, as the two protagonists draw opposite conclusions from the existence of uncertainty. Ultimately, it may be seen as necessary to incorporate the decision philosophy into legislation pertaining to the management of these vitally important resources.

Appendix

The assertion made in Section 6.2 that the risk-averse fisherman will exert less effort than the risk-neutral fisherman will now be proved. The proof is due to Sandmo (1971), adapted by Andersen (1982).

The mathematical formulations of the optimization problems are:

$$\text{Risk-neutral:} \quad \max_{E \geq 0} E\{\pi(E, p)\} \tag{6.102}$$

$$\text{Risk-averse:} \quad \max_{E \geq 0} E\{U(\pi(E, p)\} \tag{6.103}$$

where

$$\pi(E, p) = pqXE - c(E) \tag{6.104}$$

and

$$U' > 0, \quad U'' < 0 \tag{6.105}$$

We are treating price p as a random variable with known distribution $\phi(p)$. We also assume increasing marginal effort costs, $c''(E) > 0$.

Let E_1, E_2 denote the two solutions. From (6.102) we have

$$c'(E_1) = \bar{p}qX \qquad (6.106)$$

and we wish to prove that

$$c'(E_2) < \bar{p}qX \qquad (6.107)$$

from which it follows that $E_2 < E_1$.

Differentiating in Eq. (6.103) with respect to E, we obtain

$$E\{U'(\pi)(pqX - c'(E_2))\} = 0$$

This can be written as

$$E\{U'(\pi)(p - \bar{p})qX\} = [c'(E_2) - \bar{p}qX]E\{U'(\pi)\} \qquad (6.108)$$

Now for $E = E_2$ we have

$$E\{\pi\} = \bar{p}qXE_2 - c(E_2)$$

so that

$$\pi = \pi(E_2, p) = E\{\pi\} + (p - \bar{p})qXE_2$$

Hence $p > \bar{p}$ implies $\pi(E_2, p) > E\{\pi(E_2, p)\}$, which implies $U'(\pi(E_2, p)) < U'(E\{\pi(E_2, p)\})$, and conversely. We therefore obtain

$$U'(\pi)(p - \bar{p})qXE_2 < U'(E\{\pi\})(p - \bar{p})qXE_2$$

for all $p \neq \bar{p}$. Taking expectations, we find

$$E\{U'(\pi)(p - \bar{p})qXE_2\} < 0$$

Combining this with (6.108), we finally obtain

$$C'(E_2) - \bar{p}qX < 0 \qquad \text{Q.E.D.}$$

Bibliography

Adasiak, A. 1978. The Alaskan experience with limited entry, in R. B. Rettig and J. J. C. Ginter (Eds.), *Limited Entry as a Fishery Management Tool*. Univ. Washington Press, Seattle, pp. 271–299.

Akenhead, S. A., J. Carscadden, H. Lear, G. R. Lilly, and R. Wells. 1982. Cod-capelin interactions off Northeast Newfoundland and Labrador, in M. D. Mercer (Ed.), *Multispecies Approaches to Fisheries Management Advice*. Canad. Spec. Publ. Fish. Aquat. Sci. 59:141–148.

Allen, K. R. 1973. Analysis of the stock-recruitment relation in Antarctic fin whales, in B. Parrish (Ed.), *Fish Stocks and Recruitment, Cons. Inter. Expl. Mer, Rapp. et Proc.-Verb. des Reunions* 164:132–137.

Andersen, P. 1982. Commercial fisheries under price uncertainty. *J. Environ. Econ. Manag.* 9:11–28.

Anderson, L. G. 1977. *The Economics of Fisheries Management*. Johns Hopkins Univ. Press, Baltimore.

_____ (Ed.). 1981. *Economic Analysis for Fisheries Management Plans*. Ann Arbor Science, Ann Arbor, Mich.

_____. 1984. Uncertainty in the fisheries management process. *Mar. Res. Econ.* 1:77–88.

Arrow, K. A. 1964. Optimal capital policy, the cost of capital, and myopic decision rules. *Ann. Inst. Statist. Math.* 16:21–30.

_____. 1968. Optimal capital policy with irreversible investment, in J. N. Wolfe (Ed.), *Value, Capital, and Growth: Papers in Honour of Sir John Hicks*. Edinburgh Univ. Press, Edinburgh, pp. 1–20.

270

_____. 1974. *Essays in the Theory of Risk Bearing.* North-Holland, Amsterdam.

Beddington, J. R., and C. W. Clark. 1984. Allocation problems between national and foreign fisheries with a fluctuating fish resource. *Mar. Res. Econ.* 1:137–154.

Beddington, J. R., and J. Cooke. 1983. The potential yield of previously unexploited stocks. United Nations FAO Fisheries Tech. Paper 242.

Beddington, J. R., and R. M. May, 1977. Harvesting natural populations in a randomly fluctuating environment. *Science* 197:463–465.

Beddington, J. R., C. M. K. Watts, and W. D. C. Wright. 1975. Optimal cropping of self-reproducible natural resources. *Econometrica* 43:789–802.

Bell, F. W. 1978. *Food from the Sea: The Economics and Politics of Ocean Fisheries.* Westview Press, Boulder, Colo.

Bellman, R. 1957. *Dynamic Programming.* Princeton Univ. Press, Princeton.

Berck, P. 1979. Open access and extinction. *Econometrica* 47:877–882.

Beverton, R. J. H., and S. J. Holt. 1957. *On the Dynamics of Exploited Fish Populations.* Ministry of Agriculture, Fisheries and Food (London), Fish. Invest. Ser. 2(19).

Birnie, P. 1982. IWC—a new era. *Marine Policy* 6:74–76.

Bishop, R. C., D. W. Bromley, and S. Langdon. 1981. Implementing multiobjective management of commercial fisheries: a strategy for policy-relevant research, in L. G. Anderson (Ed.), *Economic Analysis for Fisheries Management Plans.* Ann Arbor Science, Ann Arbor, Mich., pp. 197–222.

Bledsoe, L. J., J. A. Buss, and D. D. Huppert. 1974. The potential yield of the Pacific halibut industry. Univ. of Wash. Center for Quant. Sci., Seattle. Norfish Tech. Rep. No. 51.

Bockstael, N. E., and J. J. Opaluch. 1983. Discrete modelling of behavioral response under uncertainty: the case of the fishery. *J. Environ. Econ. Manag.* 10:125–137.

Botsford, L. 1981. Optimal fishery policy for size-specific, density-dependent population models. *J. Math. Biol.* 12:265–293.

Brauer, F., and A. C. Soudack. 1979. Stability regions and transition phenomena for harvested predator–prey systems. *J. Math. Biol.* 7:319–337.

Breder, C. M., Jr. 1967. On the survival value of fish schools. *Zoologica* 52:25–40.

Butterworth, D. S. 1980. The value of catch-statistics-based management techniques for heavily fished pelagic stocks, with special reference to the recent decline of the Southwest African pilchard stock. International Council for Southeastern Atlantic Fisheries (ICSEAF) Colln. Scient. Pap. (Part II), pp. 69–84.

Calkins, T. P. 1961. Measurements of population density and concentration of fishing effort for yellowfin and skipjack tuna, as indicated by catches in the Eastern Tropical Pacific Ocean. *Bull. Inter-Amer. Trop. Tuna Comm.* 4:447–492.

Canadian Department of Fisheries and Oceans. 1983. *Enterprise Allocations for the Atlantic Offshore Groundfish Fisheries.* Ottawa.

Chapman, D. G. 1964. A critical study of the Pribolof fur seal population estimates. *Fish. Bull.* 63:657–669.

Charles, A. 1983a. Optimal fisheries investment: comparative dynamics for a deterministic seasonal fishery. *Can. J. Fish. Aquat. Sci.* 40:2067–2089.

_____. 1983b. Optimal fisheries investment under uncertainty. *Can. J. Fish. Aquat. Sci.* 40:2080–2091.

_____. 1983c. Effects of parameter uncertainty and Bayesian updating on fisheries investment. Univ. of British Columbia Inst. Appl. Math. Stat. Tech. Rep. No. 83-2.

272   Bibliography

Christy, F. T., Jr. 1977. Limited access systems under the Fishery Conservation and Management Act of 1976, in L. G. Anderson (Ed.), *Economic Impacts of Extended Fishery Jurisdiction*. Ann Arbor Science, Ann Arbor. Mich. pp. 141–156.

Christy, F. T., Jr. and A. D. Scott. 1965. *The Common Wealth in Ocean Fisheries*. Johns Hopkins Univ. Press, Baltimore.

Chuma, J. L. 1981. A multiple age-class population model with delayed recruitment. M.Sc. thesis, Univ. British Columbia, Vancouver.

Ciriacy-Wantrup, S. V. 1968. *Resource Conservation: Economics and Policies*. 3rd ed. Univ. of California, Div. of Agric. Sciences, Berkeley.

Ciriacy-Wantrup, S. V., and R. C. Bishop. 1975. Common property as a concept in natural resources policy. *Nat. Res. J*. 15:714–717.

Clark, C. W. 1971. Economically optimal policies for the utilization of biologically renewable resources. *Math. Biosci*. 12:245–260.

_____. 1973a. The economics of overexploitation. *Science* 181:630–634.

_____. 1973b. Profit maximization and the extinction of animal species. *J. Polit. Econ*. 81:950–961.

_____. 1974. Possible effects of schooling on the dynamics of exploited fish populations. *J. Cons. Intern. Expl. Mer*. 36:7–14.

_____. 1975. The economics of whaling: a two-species model. In G. S. Innis (Ed.), *New Directions in the Analysis of Ecological Systems*. Simulation Councils Proc. Ser. 5(2):111–131.

_____. 1976a. *Mathematical Bioeconomics: The Optimal Management of Renewable Resources*. Wiley–Interscience, New York.

_____. 1976b. A delayed-recruitment model of population dynamics, with an application to baleen whale populations. *J. Math. Biol*. 31:381–391.

_____. 1980a. Towards a predictive model for the economic regulation of commercial fisheries. *Can. J. Fish. Aquat. Sci*. 37:1111–1129.

_____. 1980b. Restricted access to common-property fishery resources: a game-theoretic analysis, in P. T. Liu (Ed.), *Dynamic Optimization and Mathematical Economics*. Plenum, New York, pp. 117–132.

_____. 1982. Concentration profiles and the production and management of marine fisheries, in W. Eichhorn (Ed.), *Economic Theory of Natural Resources*. Physica-Verlag, Wurzburg-Wien, pp. 97–112.

_____. 1984. The effect of fishermen's quotas on expected catch rate. *Marine Res. Econ*. (in press).

Clark, C. W., A. Charles, J. R. Beddington, and M. Mangel. 1985. Optimal capacity decisions in a developing fishery. *Marine Res. Econ*. (in press).

Clark, C. W., F. H. Clarke, and G. R. Munro. 1979. The optimal exploitation of renewable resource stocks: problems of irreversible investment. *Econometrica* 47:25–49.

Clark, C. W., G. W. Edwards, and M. Friedlaender. 1973. Beverton–Holt model of a commercial fishery: optimal dynamics. *J. Fish. Res. Board Can*. 30:1629–1640.

Clark, C. W., and G. P. Kirkwood. 1979. Bioeconomic model of the Gulf of Carpentaria prawn fishery. *J. Fish. Res. Board Can*. 36:1304–1312.

_____. 1984. On uncertain renewable resource stocks: optimal harvest policies and the value of stock surveys (to appear).

Clark, C. W., and R. H. Lamberson. 1982. An economic history and analysis of pelagic whaling. *Marine Policy* 6:103–120.

Clark, C. W., and M. Mangel. 1979. Aggregation and fishery dynamics: a theoretical study of schooling and the purse seine tuna fisheries. *Fishery Bull.* 77:317–337.

_____. 1984. Flocking and foraging strategies: information in an uncertain environment. *Amer. Nat.* 123:626–641.

Clark, C. W., and G. R. Munro. 1975. The economics of fishing and modern capital theory: a simplified approach. *J. Environ. Econ. Manag.* 2:92–106.

_____. 1978. Renewable resource management and extinction. *J. Environ. Econ. Manag.* 5:198–205.

_____. 1980. Fisheries and the processing sector: some implications for management policy. *Bell J. Econ.* 11:603–616.

Clark, S. H., W. J. Overholtz, and R. C. Hennemuth. 1980. *Review and Assessment of the Georges Bank and Gulf of Maine Haddock Fishery.* US National Marine Fisheries Service, Northeast Fisheries Center, Woods Hole, Mass.

Comins, H. N., and M. P. Hassell. 1979. The dynamics of optimally foraging predators and parasitoids. *J. Animal Ecol.* 48:335–351.

Cooke, J. G. 1984. On the relationship between catch per unit effort and whale abundance. *Rep. Intern. Whaling Comm.* 34. (in press).

Copes, P. 1972. Factor rents, sole ownership, and the optimum level of fisheries exploitation. *Manchester School of Soc. and Econ. Stud.* 40:145–163.

Cranfield, H. J. 1979. The biology of the oyster, *Ostrea luteria*, and the oyster fishery of Fouveaux Strait. *Cons. Intern. Explor. Mer, Rapp. et Proc.-Verb. des Reunions* 175:44–49.

Cropper, M. L., D. R. Lee, and S. S. Pannu. 1979. The optimal extinction of a renewable natural resource. *J. Environ. Econ. Manag.* 6:341–349.

Crutchfield, J. A., and G. Pontecorvo. 1969. *The Pacific Salmon Fisheries: A Study in Irrational Conservation.* Johns Hopkins Univ. Press. Baltimore.

Cushing, D. H. 1971. The dependence of recruitment on parent stock in different groups of fishes. *J. Cons. Intern. Expl. Mer.* 33:340–362.

Cushing, D. H., and J. K. G. Harris, 1973. Stock and recruitment and the problem of density dependence. In R. B. Parrish (Ed.), *Fish Stocks and Recruitment, Cons. Intern. Expl. Mer, Rapp. et Proc.-Verb. des Reunions* 164:142–155.

Dasgupta, P. S., and G. M. Heal. 1979. *Economic Theory and Exhaustible Resources.* Cambridge Univ. Press, Cambridge.

de Groot, M. H. 1970. *Optimal Statistical Decisions.* McGraw-Hill, New York.

Deriso, R. B. 1980. Harvesting strategies and parameter estimation for an age-structured model. *Can. J. Fish. Aqua. Sci.* 37:268–282.

Doubleday, W. G., and D. Rivard (Eds.). 1981. *Bottom trawl surveys.* Can. Spec. Pub. Fish. Aquat. Sci. 58.

Emlen, J. M. 1973. *Ecology: An Evolutionary Approach.* Addison-Wesley, Reading. Mass.

Feichtinger, G. 1982. Optimal bimodal harvest policies in age-specific bioeconomic models, in G. Feichtinger and P. Kell (Eds.), *Operations Research in Progress,* Deidel, Vienna, pp. 285–299.

Feldstein, M. S. 1964. The social time-preference rate in cost benefit analysis. *Econ. J.* 74:360–379.

Feller, W. 1957. *An Introduction to Probability Theory and Its Applications,* Vol. I. Wiley, New York.

Fraser, G. A. 1978. License limitation in the British Columbia salmon fishery, in B. R. Rettig and J. J. C. Ginter (Eds.), *Limited Entry as a Fishery Management Tool*. Univ. Washington Press, Seattle, pp. 358–381.

_____. 1980. Licence limitation in the British Columbia roe herring fishery: an evaluation, in N. H. Sturgess and T. F. Meany (Eds.), *Policy and Practice in Fisheries Management*, Australian Govt. Publ. Service, Canberra, pp. 117–138.

Frye, R. 1983. Climatic change and fisheries management. *Nat. Res. J.* 23:77–96.

Getz, W. 1980. The ultimate-sustainable yield problem in nonlinear age-structured populations. *Math. Biosci.* 48:279–292.

Gittins, J. C. 1979. Bandit processes and dynamic allocation indices. *J. Roy. Statist. Soc. B* 41:148–177.

Goh, B. S. 1980. *Management and Analysis of Biological Populations*. Elsevier, Amsterdam.

Goh, B. S. and T. T. Agnew. 1978. Stability in a harvested population with delayed recruitment. *Math. Biosci.* 42:187–197.

Gordon, H. S. 1954. The economic theory of a common property resource: the fishery. *J. Polit. Econ.* 62:124–142.

Graham, M. 1952. Overfishing and optimum fishing. *Cons. Intern. Expl. Mer, Rapp. et Proc.-Verb. des Reunions* 132:72–78.

Gulland, J. A. 1956. The study of the fish populations by the analysis of commercial catches. *Cons. Intern. Expl. Mer, Rapp. et Proc.-Verb. des Reunions* 140:21–27.

_____ (Ed.). 1964a. *On the measurement of abundance of fish stocks. Cons. Intern. Expl. Mer, Rapp. et Proc.-Verb. des Reunions* 155.

_____. 1964b. The reliability of catch per unit effort as a measure of abundance of the North Sea trawl fisheries. In J. A. Gulland (Ed.), *On the Measurement of Abundance of Fish Stocks, Cons. Intern. Expl. Mer, Rapp. et Proc.-Verb. des Reunions* 155:98–102.

_____. 1966. *Manual of Sampling and Statistical Methods for Fisheries Biology*. Part I. United Nations Food and Agriculture Organization, Rome.

_____. 1968. The concept of marginal yield from exploited fish stocks. *J. Cons. Intern. Expl. Mer* 32:256–261.

_____ (Ed.). 1977. *Fish Population Dynamics*. Wiley–Interscience, New York.

_____. 1978. Fishery management: new strategies for new conditions. *Trans. Amer. Fish Soc.* 107:1–11.

Gulland, J. A., and L. K. Boerema. 1973. Scientific advice on catch levels. *Fish. Bull.* 71:325–336.

Gulland, J. A., and S. M. Garcia. 1984. Observed patterns in multispecies fisheries, in R. May (Ed.), *Exploitation of Marine Communities. Dahlem Konferenzen*. Springer-Verlag, Berlin, 155–190.

Hancock. D. A. 1980. Research for management of the rock lobster fishery of Western Australia. *Proc. 33rd Ann. Conf. Gulf and Caribb. Fish. Inst., San Jose, Costa Rica*, pp. 207–229.

Hannesson, R. 1975. Fishery dynamics: a North Atlantic cod fishery. *Can. J. Econ.* 8:151–173.

Hardin, G. 1968. The tragedy of the commons. *Science* 162:1243–1247.

Hausman, J. A. 1979. Individual discount rates and the purchase and utilization of energy-using durables. *Bell J. Econ.* 10:33–54.

Healey, M. C. 1982, Multispecies, multistock aspects of Pacific salmon management, in M. C. Mercer (Ed.), *Multispecies Approaches to Fisheries Management Advice*. Canad. Spec. Publ. Fish. Aquat. Sci. 59:119–126.

Heinsohn, C. E. 1972. A study of dugongs (*Dugong dugon*) in northeastern Queensland, Australia. *Biol. Conserv.* 4:205–213.

Hennemuth, R. C., J. E. Palmer, and R. B. E. Brown. 1980. A statistical description of recruitment in eighteen selected fish stocks. *J. Northwest Atl. Fish. Soc.* 1:101–111.

Herfindahl, O. C., and A. V. Kneese, 1974. *Economic Theory of Natural Resources.* Merrill, Columbus, Ohio.

Heyman, D. P., and M. J. Sobel. 1984. *Stochastic Models in Operations Research,* Vol. II. McGraw-Hill, New York.

Hicks, J. R. 1946. *Value and Capital.* 2nd ed. Oxdord Univ. Press, London.

Hilborn, R., and M. Ledbetter. 1979. Analysis of the British Columbia salmon purse-seine fleet: dynamics of movement. *J. Fish. Res. Board Canada* 36:384–391.

Holling, C. S. (Ed.). 1978. *Adaptive Environmental Assessment and Management.* Wiley, New York.

Holt, S. J., and L. M. Talbot. 1978. New principles for the conservation of wild living resources. *Wildlife Monog.* No. 59.

Hotelling, H. 1931. The economics of exhaustible resources. *J. Polit. Econ.* 39:137–175.

Hull, J. C., P. G. Moore, and H. Thomas. 1973. Utility and its measurement. *J. Roy. Stat. Soc. A* 136(2):226–247.

Huppert, D. D. 1982. California's management programmes for the herring roe and abalone fisheries, in N. H. Sturgess and T. F. Meany (Eds.), *Policy and Practice in Fisheries Management,* Australian Govt. Publ. Service, Canberra, pp. 89–116.

Instițut del Mar del Peru. 1974. Report of the Fourth Session of the Panel of Experts on Stock Assessment of the Peruvian Anchoveta,Callao, Peru. Boletin Vol. 2, No. 10.

Inter-American Tropical Tuna Commission. 1975. *Annual Report.* La Jolla, Calif.

Jones, R. 1982. Species interactions in the North Sea, in M. C. Mercer (Ed.), *Multispecies Approaches to Fisheries Management Advice.* Can. Spec. Publ. Fish. Aquat. Sci. 59, pp. 48–63.

Kamien, M. I., and N. L. Schwartz. 1977. Optimal accumulation and durable goods production. *Zeitschr. fur Nationalokonomie* 37:25–43.

_____. 1981. *Dynamic Optimization: The Calculus of Variations and Optimal Control in Economics and Management.* North Holland, New York.

Kaufman, G. M., and H. Thomas. 1977. *Modern Decision Analysis.* Penguin, Harmondsworth, U.K.

Keyfitz, N. 1968. *Introduction to the Mathematics of Population.* Addison-Wesley, Reading, Mass.

Khoo Khay Huat. 1980. Implementation of regulations for domestic fishermen, in F. T. Christy, Jr. (Ed.), *Law of the Sea: Problems of Conflict and Management of Fisheries in Southeast Asia.* Intern. Center Living Aquat. Res. Management, Manila, Conf. Proc. 2, pp. 49–59.

Kirkwood, G. P. 1982. Simple models for multispecies fisheries, in D. Pauly and G. I. Murphy (Eds.), *Theory and Management of Tropical Fisheries.* Intern. Center Living Aquat. Res. Management, Manila, pp. 83–98.

Koopman, B. O. 1980. *Search and Screening: General Principles with Historical Applications.* Pergamon, Elmsford, N.Y.

Laevastu, T., F. Favorite, and H. A. Larkins. 1982. Resource assessment and evaluation of the dynamics of the fishery resources in the northeastern Pacific with numerical ecosystem models, in M. C. Mercer (Ed.), *Multispecies Approaches to Fisheries*

Management Advice. Canad. Spec. Publ. Fish. Aquat. Sci. 59, pp. 70–81.

Larkin, P. A. 1966. Exploitation in a type of predator–prey relationship. *J. Fish. Res. Board Canada* 23:349–356.

_____. 1977. An epitaph for the concept of Maximum Sustainable Yield. *Trans. Amer. Fish. Soc.* 106:1–11.

_____. 1978. Fisheries management—an essay for ecologists. *Ann. Rev. Ecol. Syst.* 9:57–73.

Larkin, P. A., and W. Gazey. 1982. Applications of ecological simulation models to management of tropical multispecies fisheries, in D. Pauly and G. I. Murphy (Eds.), *Theory and Management of Tropical Fisheries*. Intern. Center Living Aquat. Res. Management, Manila, pp. 123–140.

Laws, R. M. 1962. Some effects of whaling on the southern stocks of baleen whales, in E. D. Le Cren and M. W. Holdgate (Eds.), *The Exploitation of Natural Animal Populations*, Blackwell, Oxford, pp. 137–158.

Leaman, B. M. 1981. A brief review of survey methodology with regard to groundfish stock assessment, in W. G. Doubleday and D. Rivard (Eds.), *Bottom Trawl Surveys*. Can. Spec. Publ. Fish. Aquat. Sci. 58:113–123.

Levhari, D., R. Michener, and L. J. Mirman. 1981. Dynamic programming models of fishing: competition. *Amer. Econ. Rev.* 71:649–661.

Levhari, D., and L. J. Mirman. 1980. The great fish war: an example using a dynamic Cournot–Nash solution. *Bell J. Econ.* 11:322–344.

Levin, S. A., and C. P. Goodyear. 1980. Analysis of an age-structured fishery model. *J. Math. Biol.* 9:245–274.

Loasby, B. J. 1976. *Choice, Complexity, and Ignorance—An Enquiry into Economic Theory and the Practice of Decision-making*. Cambridge Univ. Press, Cambridge.

Luce, R. D., and H. Raiffa. 1957. *Games and Decisions: Introduction and Critical Survey*. Wiley, New York.

Ludwig, D. A. 1974. *Stochastic Population Theories: Lecture Notes in Biomathematics*, Vol. 3. Springer-Verlag, Berlin.

_____. 1982. Harvesting strategies for a randomly fluctuating population. *J. Cons. Intern. Expl. Mer*, 39:168–174.

Ludwig, D. A., and R. Hilborn. 1982. Management of possibly overexploited stocks on the basis of catch and effort data. Univ. British Columbia, Vancouver, Inst. Appl. Math. Stat. Tech. Rep. No. 82-3.

Ludwig, D. A., and C. J. Walters. 1981. Measurement errors and uncertainty in parameter estimates for stock and recruitment. *Can. J. Fish. Aquat. Sci.* 38:711–720.

_____. 1982. Optimal harvesting with imprecise parameter estimates. *Ecol. Modelling* 14:273–292.

McCall, A. D. 1976. Density dependence of catchability coefficient in the California Pacific sardine, *Sardinops sagax caerula*, purse seine fishery. Calif. Coop. Ocean. Fish. Invest. Rep. 18, pp. 136–148.

McKellar, N. B. 1982. The political economy of fisheries management in the Northeast Atlantic, in N. B. Sturgess and T. F. Meany, (Eds.), *Policy and Practice in Fisheries Management*, Australian Govt. Publ. Sergice, Canberra, pp. 349–364.

McKelvey, R. W. 1981. Economic regulation of targeting behavior in a multispecies fishery. Univ. British Columbia (Vancouver). Dept. Econ. Resource Paper No. 75.

_____. 1983. The fishery in a fluctuating environment: coexistence of specialist and generalist vessels in a multipurpose fleet. *J. Environ. Econ. Manag.* 10:287–309.

_____. 1984. The dynamics of open-access exploitation of a renewable resource: the case of irreversible investment. Univ. Montana Math. Dept. Interdisc. Ser. Rep. No.24.

Majumdar, M., and T. Mitra. 1983. Dynamic optimization with a nonconvex technology: the case of a linear objective function. *Rev. Econ. Stud.* 50:143–151.

Mangel, M. 1981. Search, effort, and catch rates in fisheries. *Eur. J. Oper. Res.* 11:361–366.

Mangel, M., and R. E. Plant. 1984. Regulation and information processing in uncertain fisheries. *Marine Res. Econ.* (in press).

Marasco, R. J., and J. M. Terry. 1982. Controlling incidental catch—an economic analysis of six management options. *Marine Policy* 6:131–140.

May, R. M. 1975. *Stability and Complexity in Model Ecosystems.* 2nd ed. Monographs in Population Biology VI, Princeton.

_____. 1980. Mathematical models in whaling and fisheries management. *Lecture Notes Math. Life Sci.* 13:1–62.

_____ (Ed.). 1981. *Theoretical Ecology: Principles and Applications.* 2nd ed. Blackwell, Oxford.

_____(Ed.). 1984. *Exploitation of Marine Communities.* Dahlem Konferenzen. Springer-Verlag, Berlin, in press.

May, R. M., J. R. Beddington, C. W. Clark, S. J. Holt, and R. M. Laws, 1979. Management of multispecies fisheries. *Science* 205:267–277.

May, R. M., J. R. Beddington, J. W. Horwood, and J. A.Shepherd. 1978. Exploiting natural populations in an uncertain world. *Math. Biosci.* 2:219–252.

Meany, T. F. 1978. Restricted entry in Australian fisheries, in R. B. Rettig and J. J. C. Ginter (Eds.), *Limited Entry as a Fishery Management Tool.* Univ. Washington Press, Seattle, pp. 391–415.

Mendelssohn, R. 1980a. A systematic approach to determining mean-variance tradeoffs when managing randomly varying populations. *Math. Biosci.* 50:75–84.

_____. 1980b. Using Markov decision models and related techniques for purposes other than simple optimization: analyzing the consequences of policy alternatives on the management of salmon runs. *Fiehery Bull.* 78:35–50.

Mercer, M. C. (Ed.). 1982. *Multispecies approaches to fisheries management advice.* Canad. Spec. Publ. Fish. Aquat. Sci. 59.

Mishan, E. J. 1971. *Cost–Benefit Analysis.* Unwin, London.

Moloney, D. G., and Pearse, P. H. 1979. Quantitative rights as an instrument for regulating commercial fisheries. *J. Fish. Res. Board Can.* 36:859–866.

Munro, G. R. 1976. Application to policy problems: an example, in C. W. Clark, *Mathematical Bioeconomics: The Optimal Management of Renewable Resources.* Wiley-Interscience, New York, pp.77–86.

_____. 1979. The optimal management of transboundary renewable resources. *Can. J. Econ.* 12:355–376.

Murphy, G. I. 1977. Clupeoids, in J. Gulland (Ed.), *Fish Population Dynamics.* Wiley-Interscience, New York, pp. 283–308.

_____. 1980. Schooling and the ecology and management of marine fishes, in J. E. Bardach, J. J. Magnuson, R. C. May, and J. O. Reinhart (Eds.), *Fish Behavior and Its Use in the*

Capture and Culture of Fishes. Intern. Center Living Aquat. Res. Management, Manila, pp. 400–414.

_____. 1982. Recruitment of tropical fishes, in D. Pauly and G. I. Murphy (Eds.), *Theory and Management of Tropical Fisheries*. Intern. Center Living Aquat. Res. Management, Manila, pp. 141–148.

Nash, J. F. 1953. Two-person cooperative games. *Econometrica* 21:128–140.

Neher, P. A., and A. D. Scott. 1980. The public regulation of commercial fisheries: report to the Economic Council of Canada. Univ. British Columbia (Vancouver), Dept. of Economics.

Newman, G. G. 1984. Policy instruments available for multispecies fisheries management, in R. May (Ed.), *Exploitation of Marine Communities*. Dahlem Konferenzen, Springer-Verlag, Berlin, 313–333.

New Zealand Fishing Industry Board. 1978. The southern scallop fishery of New Zealand: a cost and earnings appraisal. Wellington.

Neyman, J. 1949. On the problem of estimating the number of schools of fish. *Univ. Calif. Publ. Stat.* 1:21–36.

Niskanen, W. A. 1971. *Bureaucracy and Representative Government*. Aldine–Atherton, Chicago.

Ocean Sciences Board, U.S. National Academy of Sciences. 1980. *Fisheries Ecology: Some Constraints That Impede Advances in Our Understanding*. Washington, D.C.

Page, T. 1976. *Conservation and Economic Efficiency*. Johns Hopkins Univ. Press, Baltimore.

Paloheimo, J. E., and Dickie, L. M. 1964. Abundance and fishing success, in J. A. Gulland (Ed.), *On the Measurement of Abundance of Fish Stocks, Cons. Intern. Expl. Mer, Rapp. et Proc.-Verb. des Reunions* 155:152–164.

Pauly, D. 1982. Dynamics of multispecies stocks. *Marine Policy* 6:72–74.

Pauly, D., and G. I. Murphy (Eds.). 1982. *Theory and Management of Tropical Fisheries*. Intern. Center Living Aquat. Res. Management, Manila.

Pearse, P. H. (Ed.). 1979. Symposium on policies for economic rationalization of commercial fisheries, *J. Fish. Res. Board Can.* 36:711–866.

_____. 1980. Property rights and regulation of commercial fisheries. *J. Bus. Admin.* 11:185–209.

_____. 1981. *Conflict and Opportunity: Towards a New Policy for Canada's Pacific Fisheries*. Govt. of Canada Comm. on Pacific Fisheries Policy, Vancouver.

Pella, J. J., and P. K. Tomlinson. 1969. A generalized stock production model. *Bull. Inter-Amer. Trop. Tuna Comm.* 13:421–496.

Peterman, R. M., W. C. Clark, and C. S. Holling. 1979. The dynamics of resilience: shifting stability domains in fish and insect systems. In R. M. Anderson, B. D. Turner, and L. R. Taylor (Eds.), *Population Dynamics*. Symp. Brit. Ecol. Soc. 20. Blackwell, Oxford, pp. 321–341.

Pielou, E. C. 1977. *Mathematical Ecology*. Wiley–Interscience, New York.

Pope, J. G. 1972. An investigation of the accuracy of virtual population analysis using cohort analysis. International Council for Northwest Atlantic Fisheries (ICNAF) Res. Bull. No. 9, pp. 65–74.

_____ (Ed.). 1975. *Measurement of fishing effort. Cons. Intern. Expl. Mer, Rapp. et Proc.-Verb. des Reunions* 168.

_____. 1980. Some consequences for fisheries management of aspects of the behavior of pelagic fish, in A. Saville (Ed.), *The Assessment and Management of Pelagic Fish Stocks, Cons. Intern. Expl. Mer, Rapp. et Proc.-Verb. des Reunions,* 155:466–476.

Popper, K. R. 1979. *Objective Knowledge: An Evolutionary Approach.* Rev. ed. Oxford Univ. Press, Oxford.

Pounder, J. R. and T. D. Rogers. 1980. The geometry of chaos: dynamics of a nonlinear second-order difference equation. *Bull. Math. Biol.* 42:551–597.

Raiffa, H. 1968. *Decision Analysis: Introductory Lectures on Choices under Uncertainty.* Addison-Wesley, Reading. Mass.

Ramsay, F. P. 1928. A mathematical theory of saving. *Econ. J.* 38:543–559.

Reed, W. J. 1974. A stochastic model for the economic management of a renewable animal resource. *Math. Biosci.* 22:313–337.

_____. 1979. Optimal escapement levels in stochastic and deterministic harvesting models. *J. Environ. Econ. Manag.* 6:350–363.

_____. 1980. Optimum age-specific harvesting in a nonlinear population model. *Biometrica* 36:579–593.

_____. 1981. Effects of environmental variability as they pertain to fisheries management, in K. B. Haley (Ed.), *Applied Operations Research in Fishing.* NATO Conference Series II—Systems Science, Vol. 10. Plenum, New York, pp. 69–80.

_____. 1983. Recruitment variability and age structure in harvested animal populations. *Math. Biosci.* 65:239–268.

Rettig; R. B., and J. J. C. Ginter (Eds.). 1978. *Limited Entry as a Fishery Management Tool.* Univ. Washington Press, Seattle.

Ricker, W. E. 1954. Stock and recruitment. *J. Fish. Res. Board Can.* 11:559–563.

_____. 1958. *Handbook of Computations for Biological Statistics of Fish Populations.* Fish. Res. Board of Canada, Ottawa.

Rothschild, B. J. 1977. Fishing effort, in J. A. Gulland (Ed.), *Fish Population Dynamics.* Wiley–Interscience, New York, pp. 96–115.

Ryan, P. J. 1981. The Northern prawn fishery: a report of an economic survey. Australia Dept. of Primary Industry, Canberra, Fisheries Report No. 32.

Samuelson, P. A. 1980. *Economics.* 11th ed. McGraw-Hill, New York.

Sandmo, A. 1931. On the theory of the competitive firm under price uncertainty. *Amer. Econ. Rev.* 61:65–73.

Saville, A. (Ed.). 1980. *The assessment and management of pelagic fish stocks, Cons. Intern. Expl. Mer, Rapp. et Proc.-Verb. des Reunions* 177.

Saville, A., and R. S. Bailey. 1980. The assessment and management of the herring stocks in the North Sea and to the West of Scotland, in A. Saville (Ed.), *The Assessment and Management of Pelagic Fish Stocks, Cons. Intern. Expl. Mer, Rapp. et Proc.-Verb. des Reunions* 177:112–142.

Schaefer, M. B. 1954. Some aspects of the dynamics of populations important to the management of commercial marine fisheries. *Bull. Inter-Amer. Trop. Tuna Comm.* 1:25–56.

_____. 1967. Fishery dynamics and the present status of the yellowfin tuna population in the Eastern Pacific Ocean. *Bull. Inter-Amer. Trop. Tuna Comm.* 12(3).

Schlaifer, R. O. 1969. *Analysis of Decisions under Uncertainty,* McGraw-Hill, New York.

Schnute, J. 1977. Improved estimates from the Schaefer production model: theoretical considerations. *J. Fish. Res. Board Can.* 34:583–603.

Scott, A. D. 1955. The fishery: the objectives of sole ownership. *J. Polit. Econ.* 63:116–124.

_____. 1977. Commentary, in L. G. Anderson (Ed.), *Economic Impacts of Extended Fisheries Jurisdiction.* Ann Arbor Science, Ann Arbor, Mich. pp. 409–414.

_____. 1982.Property rights in natural resources. Univ. British Columbia (Vancouver) Program in Nat. Res. Econ. Conf. Proc.

Shaw, E. 1970. Schooling in fishes: critique and review, in L. R. Aronson (Ed.), *Development and Evolution of Behavior.* Freeman, San Francisco, pp. 452–480.

Shepherd, J. G. 1982a. A versatile new stock-recruitment relationship for fisheries, and the construction of sustainable yield curves. *J. Cons. Intern. Expl. Mer* 40:67–75.

_____. 1982b. A family of general production curves for exploited populations. *Math. Biosci.* 59:79–93.

Silvert, W. 1973. The economics of overfishing. *Trans. Amer. Fish. Soc.* 106:121–130.

Simpson, A. C. 1982. A review of the database on tropical multispecies stocks in the Southeast Asian region, in D. Pauly and G. I. Murphy (Eds.), *Theory and Management of Tropical Fisheries.* Intern. Center Living Aquat. Res. Management, Manila, pp. 5–32.

Sindermann, C. J. 1979. Status of the Northwestern Atlantic herring stocks of concern to the United States. U.S. N.O.A.A. National Marine Fisheries Service, Northeast Fisheries Center, Highlands, N.J., Tech. Services Rel. No. 23.

Sinn, H. W. 1982. The economic theory of species extinction: comment on Smith. *J. Environ. Econ. Manag.* 9:194–198.

Sissenwine, M. P. 1984. The uncertain environment of fish harvesters, fishery scientists, and fishery managers. *Mar. Res. Econ.* 1:1–30.

Sissenwine, M. P., and J. E. Kirkley. 1982. Practical aspects and limitations of fishery management techniques. *Marine Policy* 6:43–58.

Smith, V. L. 1969. On models of commercial fishing. *J. Polit. Econ.* 77:181–198.

_____. 1977. Control theory applied to natural and environmental resources: an exposition *J. Environ. Econ. Manag.* 4:1–24.

Sobel, M. J. 1982. Stochastic fishery games with myopic equilibria, in L. J. Mirman and D. F. Spulber (Eds.), *Essays in the Economics of Renewable Resources.* North Holland, Amsterdam, pp. 259–268.

Solow, R. M. 1974. The economics of resources or the resources of economics. *Amer. Econ. Rev.* 64:1–14.

Spence, M., and D. Starrett. 1975. Most rapid approach paths in accumulation problems. *Intern. Econ. Rev.* 16:388–403.

Spulber, D. F. 1982. Adaptive harvesting of a renewable resource and stable equilibrium, in L. J. Mirman and D. F. Spulber (Eds.), *Essays in the Economics of Renewable Resources.* North Holland, Amsterdam, pp. 117–139.

Steele, J. H. 1974. *The Structure of Marine Ecosystems.* Harvard Univ. Press, Cambridge.

Sturgess, N. H., and T. F. Meany. 1982. *Policy and Practice in Fisheries Management.* Australian Dept. of Primary Industry, Canberra.

Sutinen, J. G. 1979. Fisheries renumeration systems and implications for fisheries development. *Scott. J. Polit. Econ.* 26:147–162.

Swierzbinski, J. E. 1981. Bioeconomic models of the effects of uncertainty on the economic behavior, performance, and management of marine fisheries. Ph.D. thesis in applied mathematics, Harvard Univ.

Swierzbinski, J., G. Marchesseault, L. Goudreau, W. Bossert, and K. Cain. 1980. Entry in the Northwest Atlantic herring fishery: empirical evidence for sticky equilibria. Harvard University (unpublished).

Talhelm, D. R. 1978. Limited entry in Michigan fisheries, in R. B. Rettig and J. J. C. Ginter (Eds.), *Limited Entry as a Fishery Management Tool*. Univ. Washington Press, Seattle, pp. 300–316.

Treschev, A. I. 1975. Fishing unit measures, in J. A. Pope (Ed.), *Measurement of Fishing Effort, Cons. Intern. Expl. Mer, Rapp. et Proc.-Verb. des Reunions*, 155:54–57.

Ulltang, Ø. H. 1980. Factors affecting the reaction of pelagic fish stocks to exploitation and requiring a new approach to assessment and management, in A. Saville (Ed.), *The Assessment and Management of Pelagic Fish Stocks, Cons. Intern. Expl. Mer, Rapp. et Proc. Verb. des Reunions* 177:489–504.

United Nations Food and Agriculture Organization. 1979. *Mammals in the Sea*, Vol. I, Rome.

_____. 1980. Report of the ACMRR Working Party on the Scientific Basis of Determining Management Measures. FAO Fisheries Report No. 236. Rome.

Ursin, E. 1982. Multispecies fish stock and yield assessment in ICES, In M. C. Mercer (Ed.), *Multispecies Approaches to Fisheries Management Advice, Can. Spec. Publ. Fish. Aquat. Sci.* 59:39–47.

von Neumann, J., and O. Margenstern, 1947. *The Theory of Games and Economic Behavior*. Princeton Univ. Press, Princeton.

Walters, C. J. 1975. Optimal harvest strategies for salmon in relation to environmental variability and uncertain production parameters. *J. Fish. Res. Board Can.* 32:1777–1784.

_____. 1981. Optimum escapements in the face of alternative recruitment hypotheses. *Can. J. Fish. Aquat. Sci.* 38:678–689.

_____. 1984. Methods of managing fisheries under uncertainty, in R. M. May (Ed.), *Exploitation of Marine Communities*. Dahlem Konferenzen, Springer-Verlag, Berlin, 263–274.

Walters, C. J., and R. Hilborn. 1976. Adaptive control of fishing systems. *J. Fish. Res. Board Can.* 33:145–159.

Weitzman, M. L. 1974. Prices vs. quantities. *Rev. Econ. Stud.* 41:477–491.

Wilen, J. E. 1969. Common property resources and the dynamics of overexploitation: the case of the North Pacific fur seal. Univ. British Columbia (Vancouver), Dept. Econ. Res. Paper No. 3.

Wilson, J. A. 1980. Adaptation to uncertainty and small numbers exchange: the New England fresh fish market. *Bell J. Econ.* 11:603–616.

_____. 1982. The economical management of multispecies fisheries. *Land Econ.* 58:417–434.

Wright, S. 1981. Contemporary Pacific salmon fisheries management. *N. Amer. J. Fish. Manag.* 1:29–40.

Zeeman, E. C. 1974. Levels of structure in catastrophe theory. *Proc. Int. Congress Math., Vancouver, Canada*, pp. 533–546.

Glossary
of Symbols

Because of the large number of models discussed in this book, it has not been possible to maintain a strict one-to-one relationship between mathematical symbols and their meanings. Every attempt has been made, however, to maintain a notation that is internally consistent as well as consistent with standard usage in biology and economics.

The following lists the most common uses of the symbols. Note, however, that these symbols are sometimes used with alternative meanings when there is no danger of ambiguity; thus a, b, k, n, and so on are often employed in the generic role of parameters. (An asterisk in the right hand column signifies symbols used throughout the book.)

Symbol	Meaning	Typical unit	Where used
a	Screening rate	m^3 per SFU hr	2.1
B_t	Fish stock biomass	tonnes	1.3
c	Cost of effort	\$ per SFU year	*

Symbol	Meaning	Typical unit	Where used
$c(x)$	Unit cost of fishing	\$ per tonne	*
c_f	Fixed cost	\$ per SFU	1.3
c_k	Cost of capacity	\$ per SFU	3.3
c_S	Sale price of capacity	\$ per SFU	3.3
C_t	Catch rate	tonnes per year	*
$C(E)$	Cost-of-effort function	\$ per year	*
e	Fishing effort	m^3 per hr	2.1
e	Demand elasticity	per \$	3.2
E	Fishing effort	standardized fishing units (SFU)	*
E^*	Optimal effort	SFU	*
$E\{\ldots\}$	Expectation	—	2.4
$E\{\ldots\mid\ldots\}$	Conditional expectation	—	2.4
f	Instantaneous fishing mortality	per hr	2
$f(x)$	Probability density	—	2.4
$f(x\mid y)$	Conditional probability density	—	2.4
F	Fishing mortality	per year	*
$G(x)$	Natural growth	tonnes per year	*
H_k	Annual harvest	tonnes	3.6
H_t	Harvest rate	tonnes per year	*
$\mathcal{H}$	Hamiltonian	—	*
I_t	Investment rate	\$ per year	3.3
$J_n(\cdot)$	Value function with n periods remaining	\$	*
k	Bertalanffy growth parameter	per year	1.3
K	Carrying capacity	tonnes	*

Symbol	Meaning	Typical unit	Where used
K_t	Fleet size	SFU	3.3
m	Price of quotas	$ per tonne	4.4
M	Natural mortality rate	per year	*
n	Age of maturity (delay)	years	3.7
$n(\mu; x, \sigma^2)$	Normal probability density	—	6
N	Number of fish	—	1.3
N	Number of vessels	—	4.4
p	Price of fish	$ per tonne	*
$P(Q)$	Demand schedule	$	3.2
pr(...)	Probability of event	—	2.4
pr(...|...)	Conditional probability	—	2.4
PV	Present value	$	*
q	Catchability	per SFU year	*
Q	Supply rate	—	3.2
Q	Catch quota	tonnes per year	4.4
r	Intrinsic growth rate	per year	*
r	Minimum marginal effort cost	$ per SFU year	4
R	Recruitment (number of fish)	—	1.3
S	Escapement	tonnes	3.6
t	Time	years	*
t_0	Bertalanffy growth parameter	years	1.3
T	Time horizon	years	*
T	Season length	years	1.3
$U(\cdot)$	Utility function	$ per year	3.2
$V(\cdot)$	Value function	$	1.3
$V(\cdot)$	Value function	$ per year	A1

Symbol	Meaning	Typical unit	Where used
w_t	Weight of fish of age t	kg	1.3
w_∞	Asymptotic weight	kg	1.3
X	Fish stock biomass	tonnes	*
X^*	Optimal biomass	tonnes	*
$\bar{X}$	Bionomic equilibrium	tonnes	*
Y	Fish stock biomass	tonnes	5.1
$Y(F, a_c)$	Yield function	tonnes	3.7
Z	Total mortality rate	per year	*
$Z(X)$	Resource stock value function	$	*
Z_k	Random factor	—	6.1
α	Gamma distribution parameter	—	2.4, 2.6
α_t	Discount factor	—	3.1, 3.6
γ	Depreciation rate	per year	1.2, 3.3
$\gamma(x; \nu, \alpha)$	Gamma distribution	—	2.4, 2.6
$\Gamma(x)$	Gamma function	—	*
δ	Discount rate	per year	*
ϵ	Selectivity	—	2.1
λ	Poisson search rate	per hr	2.4, 2.6
λ	Shadow price (adjoint variable)	$ per tonne	*
μ	Mean of a distribution	—	6
ν	Gamma distribution parameter	—	2.4, 2.6
π	Net revenue flow	$ per year	*
$\pi(\cdot)$	Prior distribution	—	6
ρ	Density of fish	kg/m^3	2
$\rho(X)$	Concentration profile	kg/m^3	2

Symbol	Meaning	Typical unit	Where used
σ_x^2	Variance of x	—	*
σ	Survival rate	—	3.7
τ	Tax rate	$ per tonne	4.4
θ	Effort scale factor	m^3 per SFU hr	2.2
Ω	Sustained-yield set		5.3

Index